Extending Creativity with Adobe Firefly

Create striking visuals, add text effects, and edit design
elements faster with text prompts

Rollan Bañez

Extending Creativity with Adobe Firefly

Group Product Manager: Niranjan Naikwadi

Publishing Product Manager: Nitin Nainani

Book Project Manager: Aparna Nair

Senior Content Development Editor: Shreya Moharir

Technical Editor: Rahul Limbachiya

Copy Editor: Safis Editing

Indexer: Rekha Nair

Proofreader: Shreya Moharir

Production Designer: Jyoti Kadam

Senior DevRel Marketing Coordinator: Vinishka Kalra

First published: June 2024

Production reference: 1080524

Published by Packt Publishing Ltd.

Grosvenor House

11 St Paul's Square

Birmingham

B3 1RB, UK

ISBN 978-1-83508-428-1

www.packtpub.com

I want to thank the people who have been fundamental to each of my successes and inspired me to be the best creative steward. I am grateful for my family (Jacquelene, Zoe, and Beatrice) and the educators, students, and creative professionals from the Creative Nation Academy community who always support every event, both in person and virtually.

I am dedicating this book to my late grandparents (Reynaldo and Trinidad Manarin) and the Tan family (Margie and Gregory). I am eternally grateful for your benevolence at the early stages of my life.

This book has been such a milestone for me together with the creative communities that I represent in the Philippines (Graphic Artist Philippines).

To all Filipino designers whose goal is to make the world a better place for future generations to come – keep the flag flying and be proud of it!

Mabuhay ang malikhaing Filipino!

(Long live the creative Filipinos!)

Contributors

About the author

Rollan Bañez is an Adobe Expert with over two decades of experience as an IT professional, educator, and creative designer. He's the only Adobe Community Expert, Express Ambassador, Certified Professional, Partner by Design, and Adobe Creative Educator Innovator with multiple Adobe Certified Expert/Instructor qualifications, with a growing 100+ active credentials in Asia.

Rollan onboarded 35,000+ Filipinos to Adobe Education Exchange, certified 10,000+ Adobe Creative Educator L1 qualifications, and has 26,000+ growing YouTube subscribers. His unwavering commitment to fostering innovation in education has reached 1 million viewers (2005-2023) from his lectures to almost every university, local and international. He has been an active community moderator/admin/builder of creative groups in the Philippines since 2014.

Aside from being active in Adobe, he has a CompTIA Certified Technical Trainer qualification and he is a public speaker for design conventions, focusing on artificial intelligence, and generative AI, focusing on visual imagery, a multi-ambassador of EdTech tools, such as Microsoft, Wakelet, Curipod, Apple, Screencast, Kahoot, Kami, Book Creator, Mote, Edpuzzle, and a member of the Content Authenticity Initiative.

About the reviewer

Jemmarie Bocalbos is a dynamic creative professional excelling as a multimedia artist, digital marketer, and Creative Nation Academy certified specialist. With roles as an Adobe Certified Professional, Adobe Creative Insider Ambassador, and Express Ambassador, she serves as a seasoned leader for creative teams. She embodies innovation and leadership, and embraces change, believing that the future holds vast human potential. Her various career experiences have honed her adaptability and effectiveness across various creative domains. Grateful for mentors such as Rollan Bañez, Jemmarie is driven by a passion to share her journey. Supporting the Adobe Firefly book project has brought her immense joy and fulfillment.

Table of Contents

3

Editing Images Faster using Generative Fill   49

4

Creating Stylistic Text Effects in Adobe Express   69

5

Exploring Color Options with Generative Recolor   85

Part 2: Extending your Creative Workflow

6

7

8

Preface

This book will guide you in toward an in-depth understanding of how to use Adobe Firefly to create visual elements, such as images, and text effects, edit photos, and explore color ideas by using prompts to extend your capabilities in using generative AI for developing conceptual art and design, creating assets much faster without needing to learn complex procedures.

Who this book is for

This book is for anyone who wants to harness the power of generative AI imagery, whether you are a beginner designer, educator, or a seasoned creative professional who wants to explore how you can produce images and design elements that will enable you to upskill your visual communication skills, visual thinking, and leverage Adobe's promising technology, preparing you to future-proof your career in this era of rapidly evolving, democratized creativity.

What this book covers

Chapter 1, *Getting Started with Adobe Firefly*, introduces Adobe Firefly together with existing models that will enable you to understand how it works, how to access it using an Adobe ID, get to know the best plans for subscribing, manage how generative credits are being used, and tag what you create with Content Credentials.

Chapter 2, *Using the Text to image Module*, takes a deep dive into what prompting is, how it works, using suggestions, dealing with important settings such as image aspect ratio, content type, applying styles, and how you can share, download, and extend your workflow to Adobe Express directly in Firefly.

Chapter 3, *Editing Images Faster using Generative Fill*, explores how we can edit images faster using Generative Fill, and we will be generating images on top of existing pixels, removing unnecessary ones, changing brush settings, downloading, and even handing them off for further processing in Adobe Express.

Chapter 4, *Creating Stylistic Text Effects in Adobe Express,* shows us how to create and customize stylistic text effects and explores every possible setting we can use to make unique and great-looking outputs, exclusively in Adobe Express.

Chapter 5, *Exploring Color Options with Generative Recolor*, shows us how to generate color variations based on custom prompts, explore with ready-made sample prompts, check out color harmonies, and download images, all inside Adobe Illustrator.

Chapter 6, *Accessing Adobe Firefly in Photoshop and Illustrator*, shows us how to apply what we have learned in the Firefly web browser to Creative Cloud desktop apps, such as Photoshop and Illustrator, while using it within Adobe Express and Adobe Stock.

Chapter 7, *Accessing Adobe Firefly in Adobe Express and Adobe Stock*, shows us how we can use Firefly features such as Text to Image, Generative Fill, and Text to Template within Adobe Express, as well as Text to image in Adobe Stock.

Chapter 8, *Beyond Firefly*, shows us how to access additional resources such as Substance 3D Firefly features, check what upcoming features coming to Adobe Firefly, learn techniques such as captioning images and example prompts, how to get help and community links, together with tips for learning more about Adobe Firefly.

To get the most out of this book

Basic knowledge of computing operations such as logging in to web services, visual thinking skills, and beginner-level prompting of words with an appreciation of art and design; basic knowledge of using apps such as Photoshop, Illustrator, and Express is an advantage.

Software/hardware covered in the book	Operating system requirements
Adobe ID for Adobe Firefly and Adobe Express	Windows or macOS with a minimum of 4 GB RAM
Updated Google Chrome, Microsoft Edge, or Mozilla Firefox web browser with JavaScript enabled	
Adobe Photoshop 2024	
Adobe Illustrator 2024	
Adobe Substance 3D Stager	
Adobe Substance 3D Sampler	

Supported file Formats are JPG, PNG, WebP (Generative Fill) SVG - Generative Recolor, JPG, PNG, WebP, and HEIC (Text to Generative Match).

If you are using the digital version of this book, we advise you to type the code yourself or access the code from the book's GitHub repository (a link is available in the next section). Doing so will help you avoid any potential errors related to the copying and pasting of code.

Download the color images

We also provide a PDF file that has color images of screenshots and diagrams used in the book. You can download it here: `https://static.packt-cdn.com/downloads/9781835084281_ColorImages.pdf`.

Conventions used

There are a number of text conventions used throughout this book.

Prompts appear like this:

```
Rainbow colored splash of paint
pixel art with retro color palette
shiny black and gold liquid drip
```

Bold: Indicates a new term, an important word, or words that you see onscreen. For instance, words in menus or dialog boxes appear in **bold**. Here is an example: "Click on the **Generate** button in the **Generative fill** section, as seen in *Figure 3.2.*"

> Tips or important notes
> Appear like this.

Get in touch

Feedback from our readers is always welcome.

General feedback: If you have questions about any aspect of this book, email us at customercare@packtpub.com and mention the book title in the subject of your message.

Errata: Although we have taken every care to ensure the accuracy of our content, mistakes do happen. If you have found a mistake in this book, we would be grateful if you would report this to us. Please visit www.packtpub.com/support/errata and fill in the form.

Piracy: If you come across any illegal copies of our works in any form on the internet, we would be grateful if you would provide us with the location address or website name. Please contact us at copyright@packtpub.com with a link to the material.

If you are interested in becoming an author: If there is a topic that you have expertise in and you are interested in either writing or contributing to a book, please visit authors.packtpub.com.

Share Your Thoughts

Once you've read *Extending Creativity with Adobe Firefly*, we'd love to hear your thoughts! Scan the QR code below to go straight to the Amazon review page for this book and share your feedback.

https://packt.link/r/1-835-08428-1

Your review is important to us and the tech community and will help us make sure we're delivering excellent quality content.

Download a free PDF copy of this book

Thanks for purchasing this book!

Do you like to read on the go but are unable to carry your print books everywhere?

Is your eBook purchase not compatible with the device of your choice?

Don't worry, now with every Packt book you get a DRM-free PDF version of that book at no cost.

Read anywhere, any place, on any device. Search, copy, and paste code from your favorite technical books directly into your application.

The perks don't stop there, you can get exclusive access to discounts, newsletters, and great free content in your inbox daily

Follow these simple steps to get the benefits:

1. Scan the QR code or visit the link below:

https://packt.link/free-ebook/9781835084281

2. Submit your proof of purchase.
3. That's it! We'll send your free PDF and other benefits to your email directly!

Part 1:
Meet Adobe Firefly

Welcome to *Extending Creativity with Adobe Firefly*. This book is divided into two parts: the first part deals with the fundamental modules that will enable you to understand how generative AI is slowly being implemented from the web browser into other Adobe applications that you use in your day-to-day creative activities.

This part will cover the following chapters:

- *Chapter 1, Getting Started with Adobe Firefly*
- *Chapter 2, Using the Text to image Module*
- *Chapter 3, Editing Images Faster using Generative Fill*
- *Chapter 4, Creating Stylistic Text Effects in Adobe Express*
- *Chapter 5, Exploring Color Options with Generative Recolor*

1

Getting Started with Adobe Firefly

Welcome to the world of generative AI imagery! Exciting times are coming ahead as we enable ourselves to generate unique images, use text effects, and edit images faster at speeds that we never experienced before, all while having fun in the creation process.

Let's appreciate how Adobe Firefly can offer you a design copilot or a second creative brain, which will help you brainstorm lots of content ideas, saving you time and allowing you to focus on being creative.

In this chapter, we will get acquainted with accessing Adobe Firefly in your browser and get an overview of the existing models that it has to offer. We will navigate specific sections of the interface to get more functions that will benefit you along the way, discover how generative credits work, and learn how Content Credentials will help you protect yourself and your outputs along the way.

Here is a brief overview of what we will cover:

- Getting an in-depth background on how to use the Adobe Firefly web service
- Learning how Generative Credits work together with your existing Creative Cloud subscription
- Understanding Content Credentials impact your output as final assets

Learning how to use Adobe Firefly

Adobe Firefly is a standalone web application that enables you to ideate, create, and improve your current creative workflow using generative AI. You can use prompts to easily visualize and bring life to your thoughts and translate them into visual elements, such as images and text effects, by generating color variations and replacing specific portions of existing photos to your liking. Some notable affiliations of Firefly can be seen in NVIDIA Picasso, Google Gemini, and Adobe Experience Manager, respectively. We will discuss these in *Chapter 8*.

It aims to enhance the creative process by augmenting your use of Adobe Photoshop for faster image compositing, creating vector drawings in Adobe Illustrator, full integration with Adobe Express for faster content creation, and searching for that next great photo in Adobe Stock.

It is trained using openly licensed and public domain content, making it safe for commercial use while keeping together a compensation model for creators to thrive in monetizing their contributions and protecting them in the process.

Why use Adobe Firefly?

Adobe has been putting effort into making generative AI accessible to all with the following considerations in mind:

- **Ethical considerations**: They train models to adhere to licensed content within Adobe Stock and public domain content where copyright has already expired making it commercially safe. They have also initiated the **Content Authenticity Initiative (CAI)**, pushing industry standards for the use of responsive generative AI to media, tech companies, NGOs, academics, and others. You can learn more about CAI here: `https://contentauthenticity.org/how-it-works`.

- **Functionality and ease of use**: It is tightly integrated to streamline, using a simplified approach without the need for a high level of technical knowledge to implement.

- **Updates and support**: You can expect regular updates, bug fixes, and customer support, making it dependable as a design tool for your upcoming design projects.

- **Cutting-edge features**: There are a lot of upcoming technologies coming to Creative Cloud subscribers, such as video, 3D features, and other labor-intensive tasks such as animation.

Choosing between free and premium plans

You can access Adobe Firefly in a myriad of ways, from no cost to monthly or annually, with included generative AI credits.

There are two main kinds of plans within Adobe Firefly:

- **Free plan**: There's no need for a credit card; you only need to register using an active email address. With this plan, you get 25 monthly credits.

- **Premium plan**: This requires a payment method and costs 4.99 USD on a monthly basis. With it, you get 100 monthly generative credits, Adobe Fonts Free, and no watermarks on images generated by Firefly.

My advice is to first try the free plan for users who are beginning to explore how they can incorporate generative AI into their workflow. I am using the Creative Cloud All Apps subscription plan, so everything is included.

Whether you are using the free plan or premium plan, you will have generative credits. This is because using generative AI can be process intensive and demand computation costs, and each time you hit the **Generate** button, a lot of this is happening in the background.

When you hit the **Generate** button, try to do a page refresh, or use the **Load more** button; it will automatically debit against your generative credits.

You can check the latest up-to-date information on this using this link: `https://helpx.adobe.com/firefly/using/generative-credits.html`.

Accessing Adobe Firefly in the browser

To begin, let us make sure that you have an Adobe ID account, which will enable you to access the web service, for you to get started. You can open your browser and type in the address bar `firefly.adobe.com` and press *Enter* (Windows) or *Return* (macOS).

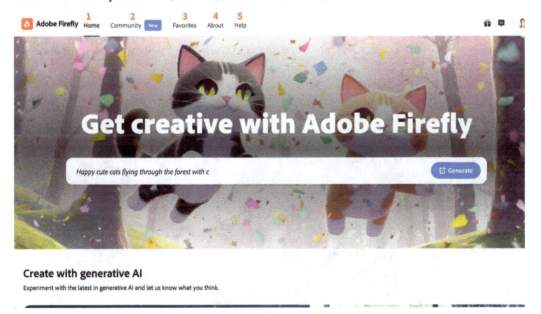

Figure 1.1 – The Adobe Firefly website home page

Let's learn what specific options we have in the menu system to get the most out of what Adobe Firefly has to offer.

Here is the description of the menu items that are labeled in the preceding screenshot:

1. **Home**: This enables you to discover all of the models available in Adobe Firefly.
2. **Community**: Here, you can discover and get inspired by community-submitted creations curated by the Firefly team for everyone to see. You can also submit your own creations and access them.

3. **Favorites**: Save your creations and you will find all of them compiled in this directory, enabling you to access them anytime with any device when you log in with your account.

4. **About**: This opens the official website for Adobe Firefly with lots of information regarding Adobe Sensei and other related topics.

5. **Help**: Get more in-depth technical information about Adobe Firefly such as data and content usage.

How to log in using an Adobe ID in Firefly

Logging into Adobe Firefly is as simple as the following steps:

1. Click the **Sign in** button in the top-right corner of the website (`firefly.adobe.com`). Signing in with an Adobe ID doesn't cost a thing – it's totally free!

2. Create your Adobe ID for you to get additional functionalities, such as saving favorites. Use an existing one that you have or create a new one. There are also options to use your Gmail, Facebook, or Apple credentials, which may be stored in your internet browser, for your ease of use.

Upon clicking your desired choice, you now have access to the Adobe Firefly website. You can now access the modules (*Figure 1.2*).

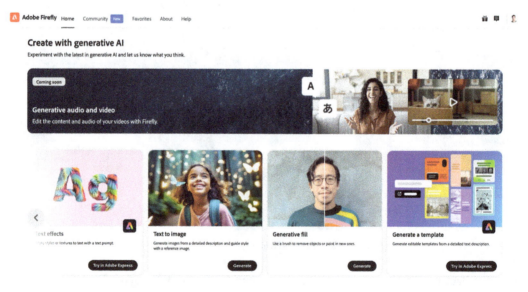

Figure 1.2 – An overview of all modules available in Adobe Firefly when you sign in

Exploring modules and menu options in Adobe Firefly

We will now discuss various input modules:

- **Text to image**: Produce compelling visuals for your social media posts, posters, flyers, and beyond effortlessly using the **Text to image** module in Adobe Firefly.

- **Generative Fill**: Edit existing pixels using a brush and seamlessly add or remove elements with ease, guided by using prompts to automatically fill the needed areas.

- **Text Effects**: Create and apply styles or textures into any text with expressive and flexible options to customize using text prompts (only available in Adobe Express).

- **Generative Recolor**: Generate color variations from vector artwork using prompts that describe any color theme. This module is available only in Adobe Illustrator at the time of writing.

- **Text to Template**: Using text prompts, you can create layouts exclusively available only in Adobe Express.

- **Text to Vector Graphic**: Make editable vector graphics using prompts exclusively inside Adobe Illustrator.

We will learn how to use these modules in the upcoming chapters, and for now, we'll shift focus to discussing some menu options that you can use with Adobe Firefly that will offer you functions that will maximize your workflow.

Using Community in Adobe Firefly

Accessing the **Community** link enables you to check on other works created by users, which you can remix and make your own. It also offers video tutorials with tips and tricks from other Firefly users.

By clicking the **View your Submissions** button on the right side, you can check whether your images are accepted by the Firefly team to be featured. I personally feel empowered by this feature.

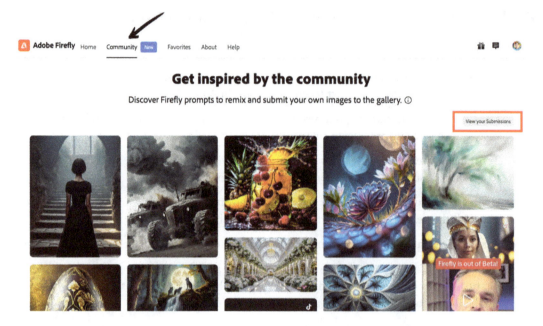

Figure 1.3 – The Community tab offers lots of inspirational links

Using Favorites in Adobe Firefly

Almost similar to the **Community** link, the **Favorites** link enables you to save your own generated images together with the prompts and settings that were used. You can have an unlimited number of assets tagged as favorites, which reside on the **Favorites** page. You can easily download them or view samples for more creative exploration that you may choose. *Figure 1.4* shows a sample **Favorites** page.

> **Tip**
>
> All of the generated images added in **Favorites** are only saved on the browser's device you are currently using; they do not appear on any other browser even if you are logged in with your Adobe ID.

Figure 1.4 – The Favorites page lets you save your work together with the prompts

Getting more information with the About page

Clicking the **About** link opens a new window, which will provide you with more information regarding Firefly, including FAQs, pricing, and a showcase of the existing models that you can use. It is a wealth of information that goes back to its root, Adobe Sensei, which has contributed a lot for creatives to save time using smart workflows.

Using Adobe Firefly Image 3 Model

The recent update to Adobe Firefly in April 2024 has given users access to a new model called **Image 3**, available in beta by default in Photoshop, InDesign, and Firefly web app.

Firefly Image 3 introduces enhancements in photorealistic quality, styling versatility, meticulous detail and precision, alongside an expanded range of options, streamlining the ideation and creation of workflow for increased productivity and efficiency.

In comparison, the left portion is Image 2 and the right portion is Image 3; you can definitely see the difference in them as seen in *Figure 1.5*.

Figure 1.5 – Image comparison of using Image 2 versus Image 3

Here are some of the reasons why Image 3 is better in generating your assets in Adobe Firefly:

- It gives a broader understanding of the world, landmarks, and cultural symbols. You also have the ability to influence image generation by using more extensive and more elaborate prompts.

- It enables an enhanced generation of individuals, especially in portraits, improves features such as skin, hair, eyes, hands, and body structures, and promotes greater diversity (e.g., by creating better hands and anatomy).

- It helps with enhanced photographic quality, incorporating high-frequency details, such as skin pores and foliage, and control over depth of field.

- There is an **Auto mode** for **Content type** that autonomously chooses between **Photo** or **Art** content types, predicting optimal photo settings for your prompt. This ensures excellent results without the need for manual adjustments.

- Lastly, the colors are more vibrant, and the dynamic range is enhanced, offering rich imagery without excessive saturation.

We will go into more detail on all of the features, such as Generative Match, Photo Settings, Prompt Suggestions, Structure and Style Reference, along with additional features that you can use as we use the Text to image module in the next chapter.

So, which Adobe Creative Cloud apps work with Firefly? Features of Firefly have been integrated into popular and industry-standard applications, such as Generative Fill and Expand in Adobe Photoshop, Text to Vector Graphic and Generative Recolor in Adobe Illustrator, content creation in Adobe Express, and search capabilities in Adobe Stock.

Use cases of using Adobe Firefly

Let us now see some use case examples with Adobe Firefly:

- **Conceptual art**: Game developers, graphic artists, and multimedia designers can create unique digital assets, providing inspiration for characters, scenes, and backgrounds. I created this image (*Figure 1.6*) using the following prompt:

```
a sea of memories and sadness, red night sky in rococo inked
drawing style.
```

Figure 1.6 – Conceptual art such as backgrounds using Firefly

- **Content creation**: Other fields can greatly facilitate the use of Adobe Firefly to create design mockups, and visual content for advertisements, social media promotions, or campaigns that they need to execute. *Figure 1.7* is an image generated by a marketing professional that translates their vision into visuals really quickly.

Figure 1.7 – Product shots can be easy to visualize by using prompts

- **Instructional material for education content**: Educators and non-teaching staff can easily participate in building educational content that can be used to facilitate improved visual aids. For instance, consider *Figure 1.8*, which demonstrates how to break down technical jargon into easy-to-understand visuals.

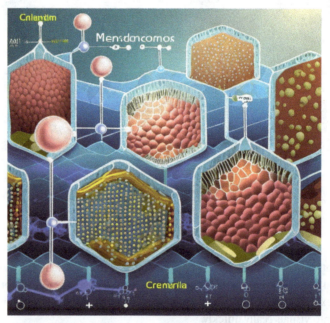

Figure 1.8 – An image illustration using the isometric view in doing technical drawings

In the next section, we will discuss how Content Credentials work and why they are important in your workflow.

Understanding Content Credentials

Content credentials are like a special kind of proof that comes with digital files. They let creators add extra information about themselves and how they made something. This extra information helps creators get more recognition for their work, connect with others online, and be clear with their audience about what they created.

These Content Credentials are part of a group of technologies called the CAI. Adobe and over 1,200 other CAI members are working together to make a standard way for people to share digital content.

This standard keeps important details about who made it and when and how it was created. Adobe also helped start another group, the **Coalition for Content Provenance and Authenticity (C2PA)**.

This group is creating a global standard for sharing this information, not just in Adobe products but across different platforms and websites. Content credentials are a way of using this standard in action.

Content credentials make it simple to share your content in a clear way. This helps you build trust with your audience by giving them more information about you and how you create. It also stops false information from spreading online.

When you use Content Credentials, you link your identity and contact details to your work. This makes it easier for people to find and get in touch with you when they come across your content on the internet.

So, how do Content Credentials work? Let's look at the following diagram:

Capture
Work with manufacturers to integrate CAI into smartphones/cameras.

Edit
Integrate CAI into editing tools, both Adobe products *and* others.

Publish
Publishing systems maintain CAI metadata throughout their platforms.

Trust
Clear *and* universal user experience reveals provenance information.

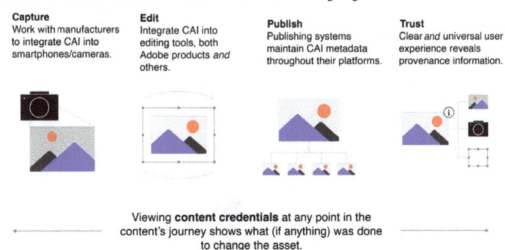

Viewing **content credentials** at any point in the content's journey shows what (if anything) was done to change the asset.

Figure 1.9 – Each step conducted in the Content Credential process

As content goes through changes or edits, it can collect more Content Credentials. This makes a history of different versions that people can check to decide how much they can trust the content.

When you save or download something, Content Credentials add extra details to it. These details are stored in a secure set of information called a Content Credential. This special information stays with the content wherever it goes, helping people understand both the content and its context. *Figure 1.10* shows an example of how Content Credentials are implemented.

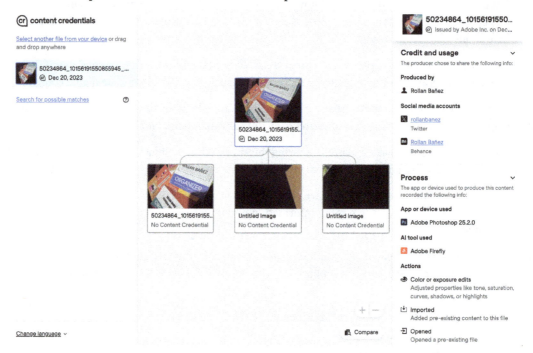

Figure 1.10 – All the needed information is being embedded into the image

How to use Content Credentials

There are two ways to use Content Credentials:

- **Attached to the file**: This makes them bigger but helps to keep your Content Credentials more private. However, these attached Content Credentials are not as strong and can be removed from your content when you share or publish it online. *Figure 1.11* shows how you can enable the attachment of Content Credentials to the file upon export of the image.

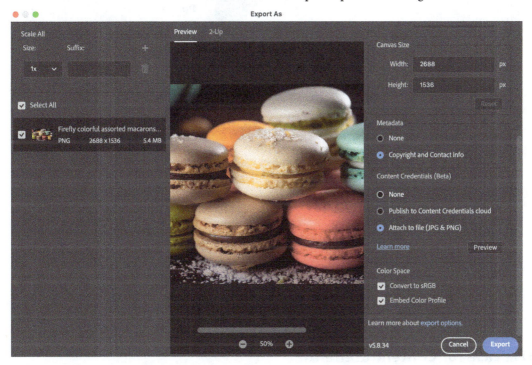

Figure 1.11 – Attaching an image to a file in Adobe Photoshop during the export

- **Via Content Credentials Cloud**: This technique makes files smaller and only verifies an image using the metadata being used when the image is exported, enabling you to verify it instantly using the Content Credentials website.

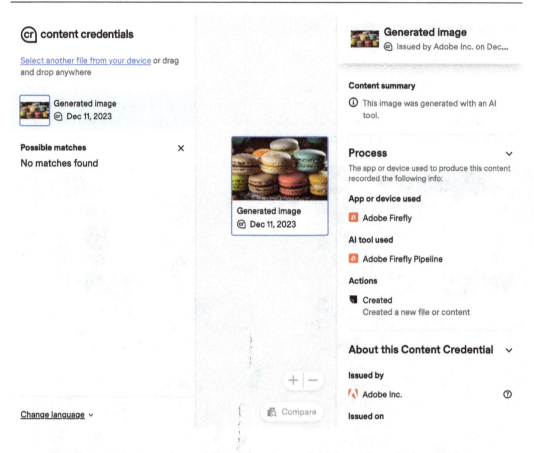

Figure 1.12 – Using the verify image function to check for additional information

Use case for Content Credentials

Content Credentials are really handy for creators. They help creators give credit and share how their work can be used. This adds an extra layer of transparency for the audience. Both casual and professional artists can use Content Credentials with the following scenarios:

- **Content accreditation**: Creators can use Content Credentials to make sure they get recognized for their work when it's shared or published. They can also say how they want others to use their content. Creators can be open about their general editing process, showing what they did to create their content without revealing all the little details. With Content Credentials, creators can also share their contact information, such as social media accounts and web addresses, without needing an Adobe account.

- **Content transparency for AI-generated work**: This enables you to get a badge with Adobe Firefly creations, letting everyone know we used generative AI tools. This helps keep things clear. Later on, other Adobe apps will also have this badge to show when generative AI was used in the creation of something.

- **Secured photojournalism**: This new way of giving credit lets anyone check reliable details about pictures. When you spot the CAI info icon (a small **i** in a circle), you can look closely at the image and see how it has changed over time. This is still in the early stage of implementation, with only a few photographers having access to this technology.

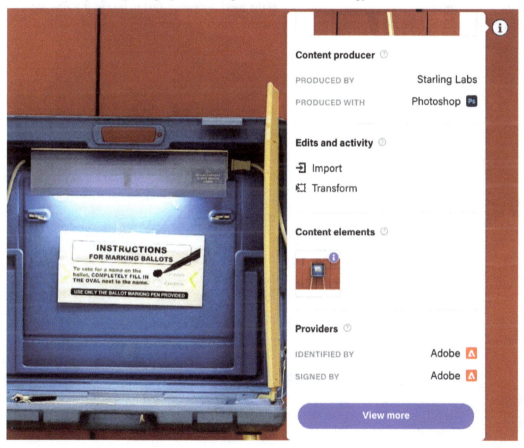

Figure 1.13 – Image information can easily be verified together with additional details

You can also see the tools they used in Photoshop to edit the picture. If you click **View more**, you can even compare different versions. This makes pictures more clear and trustworthy. Now, people reading the news or looking at social media can decide whether they want to believe what they see based on this extra information.

Summary

In this first chapter, we covered a lot of our fundamentals as we discovered the features and parts of Adobe Firefly. We discovered that specific use cases may be used to augment our creative workflows. We created our Adobe ID, which we can leverage to access more features; we covered how generative credits work and how it can impact you as a user to plan ahead if you are going to use the free or the premium plan; and lastly, we learned how we can we protect ourselves with our creations using Content Credentials.

In the next chapter, we will cover how to use one of the modules called Text to image; it is one of the most feature-rich and exciting features to learn about in Adobe Firefly.

2
Using the Text to image Module

In this chapter, we will take a deep dive into what **prompting** is. This is an exciting chapter for we will cover every possible button and explain how they work within the **Text to image module**, check out using suggestions to help you generate images better, deal with important settings such as image **aspect ratio** and **content type**, learn how to apply styles, and cover how you can share, download, and extend your workflow to Adobe Express directly in Firefly.

In this chapter, we'll cover the following topics:

- Writing effective prompts for image generation
- Generating images using the Text to image module
- Working with advanced settings of Adobe Firefly

Writing effective prompts for image generation

Describing things as thoroughly as possible can be challenging. As creators, we may not always have the actual words needed to generate images in the exact way that we want. This is where the practice of **prompt engineering** will definitely help us.

Prompt engineering is a process by which you translate pictures into words together with other parameters (properties) that you want your image to become.

Prompt engineering plays a vital role in the creation of better prompts; it is the language that the Text to image module understands. In simple terms, it is how you can ask the AI module to draw how you want it to. Let's learn how to craft a well-written text-to-image prompt.

You can follow this set of guided questions and statements to craft a structured approach to creating prompts:

1. What type of image output do you want? For example, is it a photo or a piece of art that you want to draw?
2. Are there any specific subjects that you need to be included?

Do you need a person, an animal, an object, or a majestic landscape of nature in your image?

3. Add some details to make it more precise such as lighting setup, environment, color scheme, or even a specific point of view.

    For example, we can use words such as neon light, indoor, vibrant, dim background, and overhead shot.

4. Is there a specific art style that you want to incorporate? Some suggestions we have are 3D rendering, Cubism, conceptual art, and comic-book-style art.

5. Would you like to add some photographic composition settings?

    You can get technical with it such as applying a wide angle, shooting from above, knolling, aperture setting such as f/4, shutter speed such as 1/250s, and focal length such as 50 mm.

While you can be very specific and provide so much detail in your prompt to make it fit your target output, this is just a guideline on which you can practice your prompt engineering skills.

> **Tip**
>
> Do not be afraid to experiment with settings that you can envision in your mind for you to improve. With the recent release of Adobe Firefly, you can now input prompts with support for up to 100 languages, enabling more user to use it in their native tongue.

As an activity, keep in mind and dissect the following example prompt – it provides all the needed parameters. Personally, I would recommend providing only the first three parameters as you are beginning to create your own prompts:

```
Knolling of Japanese cuisine, professional food photography,
realistic, depth of field, 4k, highly detailed, the smell permeated in
the air, Editorial Photography, Photography, Shot on 70mm lens, Depth
of Field, Bokeh, DOF, Tilt Blur, Shutter Speed 1/1000, F/22, White
Balance, 32k, Super-Resolution, white background
```

> **Tip**
>
> Try also to check your prompts with non-contradicting words. One example could be including words such as monochromatic and colorful in which the order of words will definitely matter because it will give priority to the earliest input within the structure of the prompt.

Now, let's create our first text-to-image output in Adobe Firefly!

Generating images using the Text to image module

To create wonderful images, try simple prompts first – there's no pressure and just take baby steps until we are up and running.

Let's try some prompts that you can do in Adobe Firefly:

1. Log in to the website at `https://firefly.adobe.com/` and click on the **Generate** button in the **Text to image** section as shown in *Figure 2.1*.

Text to image

Create unique images from a text prompt and apply
style presets.

Generate

Figure 2.1 – Accessing the Text to image function

2. Try this simple prompt by typing in the textbox:

    ```
    vintage camera on a table with white background
    ```

    Click the **Generate** button (see *Figure 2.2* for more guidance). We will repeatedly use this prompt throughout this chapter for you to follow along.

    Prompt

    vintage camera on a table with white background

    Figure 2.2 – Indicate the needed prompt, together with the Generate button

3.   It will create four outputs that you can choose from as seen in *Figure 2.3*. Do not worry if the image you generated does not match the figure in this book – it is supposed to build unique ones every time you hit the **Generate** button.

Figure 2.3 – Output from the example prompt provided in Step 2

4. You can click on each image to inspect the quality of the output. There are also more options such as editing and adding the image as a favorite. But for now, hover your cursor over a generated image and click the **Download** button to download the image as seen in *Figure 2.4*.

Figure 2.4 – You can finalize the image generated using the Download button

5. Before you can download the output file in JPG format, another window will ask you about Adobe's commitment to promoting transparency with regard to content generated using AI tools such as Adobe Firefly.

The file will contain Content Credentials, which lets people know that it was generated using an AI tool. You can check more information as indicated in *Figure 2.5*. If you agree to this, you can simply click the **Continue** button.

Promoting transparency in AI

Adobe is committed to promoting transparency around content generated with AI tools like Adobe Firefly.

When downloading or sharing content generated with Adobe Firefly:

(cr)   Content Credentials will be applied to let people know it was generated with AI.

🖼   The Content Credentials will note when your creation used a reference image.

Learn more about Content Credentials

☐ Don't show this again   Cancel   Continue

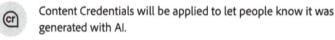

Figure 2.5 – Notification for promoting transparency with Adobe Firefly

6. This will automatically save the file in your local directory, commonly in the `Downloads` folder. Take note that the filename may also contain information about the prompt that you have supplied, making it available to everyone who has access to it.

Now, you are done with your first text-to-image creation. In the next section, we will discuss and take a deep dive into the advanced settings and controls available in the Text to image module.

Exploring features in Text to image via Adobe Firefly website

As of April 24, 2024, the default **Firefly Image 3** will be used in every Text to image generation task on the Adobe Firefly website. We will discuss several features that Adobe Firefly Image 3 has to offer.

Firefly Image 3 (Beta) has a lot to offer, such as **Photo Settings**, **Visual intensity**, **Generative Match**, **Style Reference**, and **Structured Reference** features, as seen in *Figure 2.6.*

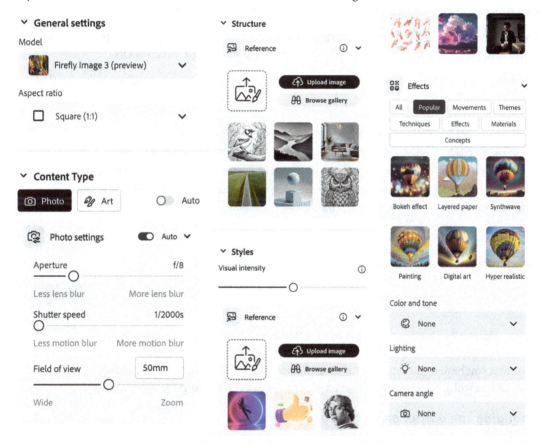

Figure 2.6 – An overview of options and controls you can use in Text to image using Firefly Image 3

This is a significant improvement that enables you to create images with accuracy and control that has never been seen before, just one year since it was first launched.

Speaking of improvements, one that will greatly benefit beginners who craft their own prompts is having prompt suggestions. This is an easy thing to enable, and we will cover in the next section.

Using prompts suggestions

There will be times when generating the precise prompt will rather feel like a tedious task, which is why the recent update of Adobe Firefly has offered **Prompt suggestions**. Think of it as autocomplete when you are typing in your text document; it gives you ideas to improve your prompt compositions.

As an activity, try it yourself with an active prompt loaded. Click at the very end of the prompt and experience prompt suggestion yourself.

You can see in *Figure 2.7* that we are using our previously created prompt, which is automatically active when you click on the textbox, given that the suggestion toggle switch is **ON**.

You still need to click the **Generate** button for you to be able to check the output of the suggestion.

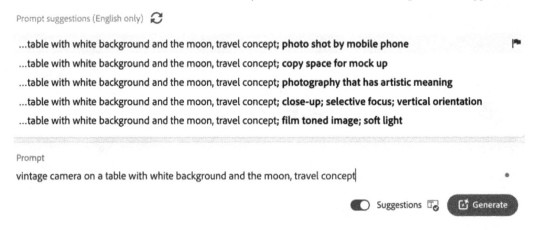

Figure 2.7 – Clicking the textbox will enable Adobe Firefly to give you prompt suggestions

Now that you have some extra help with using prompt suggestions, let's cover changing our aspect ratio, as this will greatly benefit you when generating assets for your projects.

Changing the output aspect ratio

Part of the settings you can manipulate is the aspect ratio. **Aspect ratio** deals with the specific dimension the output needs to have. You can choose from the following choices: **Square (1:1)**, **Landscape (4:3)**, **Portrait (3:4)**, and **Widescreen (16:9)**; for the most part, we only need to use the **Square** option, as seen in *Figure 2.8*.

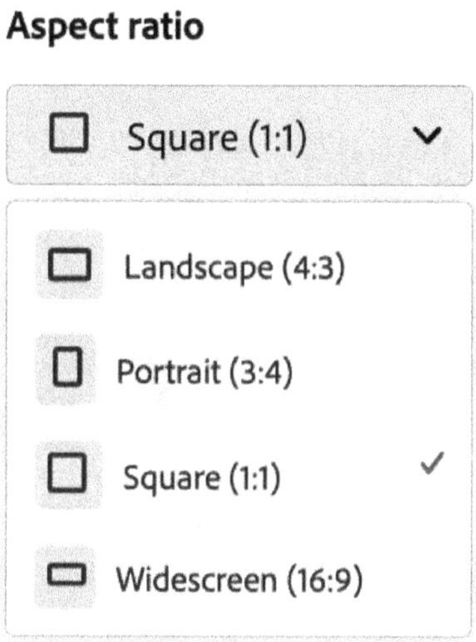

Aspect ratio

Figure 2.8 – All the available options in the aspect ratio category

As an activity, try to generate at least a set of images in **Widescreen** and **Portrait**. Those are my go-to choices in my personal workflow.

After that, let's go and make drastic changes and improve our text to image skills using **Content Type** options, coming in the next section.

Modifying content type and visual intensity

Going down in our list of settings is the content type – you can choose from **Photo** or **Art** depending on the need of your output.

Photo enables you to generate lifelike or realistic types of images while choosing the **Art** option will mostly end up being expressive and more based on the style provided in the prompt (if any). Feel free to change the prompt if you are following along.

I have used the following prompt for this example:

```
Furious blue dragon with big wings, riding in a mountain snow ready
for an attack
```

Check out the big difference between using the **Photo** and **Art** options in the results generated in *Figure 2.9*

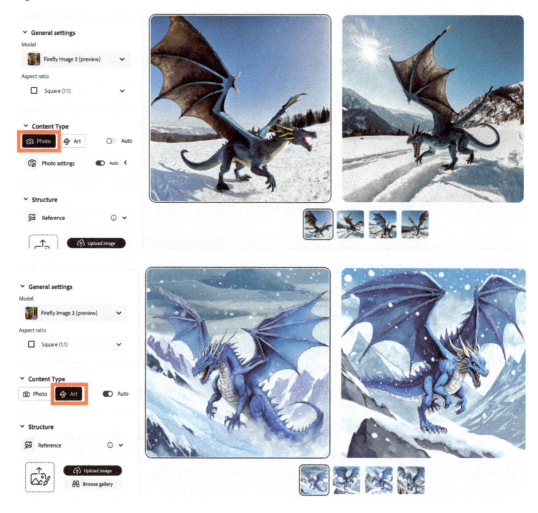

Figure 2.9 – Comparing generated output using Photo (top) and Art (bottom)

Did you notice what additional controls you enabled by using the **Photo** option? Yes, you have photo settings similar to what you have with a camera. In this case, we will be using another prompt that will suit our needs to properly demonstrate how it works. That is coming next in the following section.

Adjusting Photo settings

Photo settings allows you to dial in specific controls such as **Aperture**, **Shutter speed**, and **Field of view**. This option will only be available when you set your content type as **Photo**. Having access to this and typing your prompts together enables you to have a virtual camera that can also offer you a way to learn photography. See *Figure 2.10*, where these are all set.

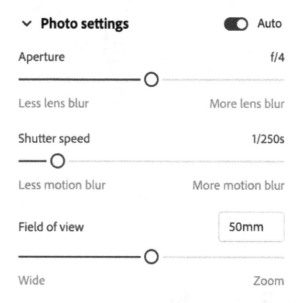

Figure 2.10 – You have Aperture, Shutter speed, and Field of view to play with

As an activity, for us to appreciate how **Photo settings** works, let's change our prompt to the following:

```
cars traveling at night in golden gate bridge
```

This is a good example. It lets us change the value of **Shutter speed** so that it produces more motion blur, which involves dragging the slider to the rightmost value. Take note that the **Auto** function is turned off in this instance. You will see in *Figure 2.11* the effect it creates in our generated output.

Figure 2.11 – Changing Shutter speed in the Photo settings function

You can also use **Photo settings** in **Auto** mode for Firefly to set the settings based on the prompt you have typed and create the necessary recommended settings for it. For example, we have used the following prompt:

```
colorful bouncing ball with a red and white background
```

Check *Figure 2.12* to see the changes in **Photo settings**.

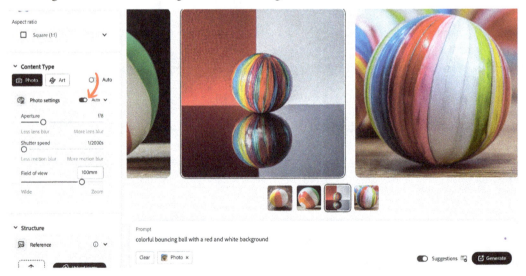

Figure 2.12 – Using Auto mode in Photo settings with the Aperture and Field of view added values

As an activity, try to generate images that will simulate some photographic techniques, such **Field of view** and **Aperture**.

I think by now you can probably see that learning these controls in Text to image can also open other avenues for us to brush up on our understanding of photography and even our appreciation of art itself.

Now, let's cover how we can make accurate positioning using Structure Reference in the next section.

Using Structure Reference

Using Text to image can be a bit of challenge, and even sometimes a bit hit and miss. This is why Structure Reference can help you become creatively consistent. But how? It enables you to guide Text to image to figure out specific points in your reference to match the outcome by focusing on the positions of elements.

Let's try this as an activity. Here are the steps of how to implement Structure Reference:

1. Change the prompt in the text box using the following:

    ```
    8 bit style, orange cat wearing large headphones and dark
    shades, with drinking straw with umbrella; creative concept;
    modern minimalistic design
    ```

2. Go to the **Structure Reference** section and click the **Browse Gallery** button.

> **Important tip**
>
> You can also use the upload image button and use your own images as Structure Reference, you just need to make sure that you own the photo just to be on the safe side of not infringing any copyright.
>
> You can also draw shapes manually with pen and paper, capture it as a digital photo, and apply it as a Structure Reference.

3. Go to the Photography category and choose the photo of a white cat staring by the window.

4. Click **Generate** at the bottom and wait for the image to generate. You will see the suggested result together with buttons in *Figure 2.13*:

Figure 2.13 – Using Structure Reference for consistent image generation using a sample image

You will notice that the positioning of the subject is being implemented in the generated images, enabling greater control of the output. This is a good way to make variations of images that you can ideate from, jumping from one art style to another.

I can definitely use this as a learning tool for students to appreciate art styles while having the same representation of the image. Speaking of styles, we can take this to another level by learning about Visual intensity in the next section.

Using Styles

Part of the Styles category is Visual intensity. This refers to how intense the detail being applied in the generated image is. *Figure 2.14* gives you a direct comparison between 0% visual intensity and 100%, with the same prompt being used on both of them.

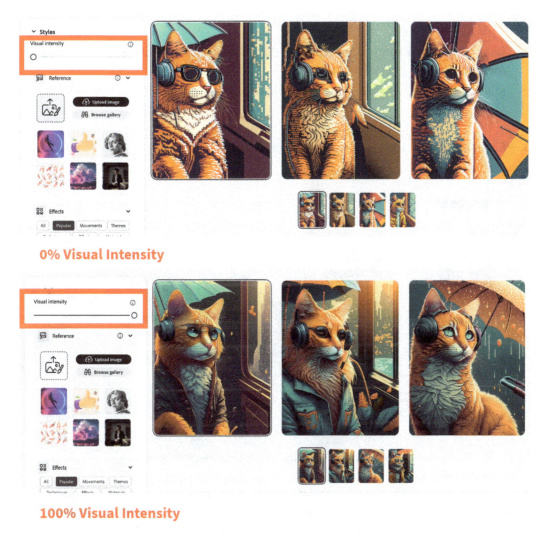

0% Visual Intensity

100% Visual Intensity

Figure 2.14 – Adjusting the output quality using the Visual intensity slider in the Styles category

Accessing Style Reference and Reference image gallery

Generative Match is a feature that was added to Adobe Firefly and released publicly in September of 2023. It enables you to include a reference image along with your text prompt. Within that feature is **Style Reference**, which enables you to generate a style based on the image. The resulting image of your prompt will fit seamlessly blending with its visual resemblance.

It aims to provide a consistent look in your image generations such as a style based on your own images (provided you confirm that you have rights to use the image) and by using the **Reference image gallery** tool that is built in Firefly.

Now, let's try to use Style Reference with the following steps:

1. Let's click the **Clear** button at the bottom of the Prompt textbox to make sure that nothing is affecting our newly generated images.

2. Now change our existing prompt and input the following:

```
vintage camera on a table with white background
```

3. Click on **Reference image gallery** on the right side of the interface as shown in *Figure 2.15*.

Figure 2.15 – Using a Style Reference with the Reference image gallery

4. Click any image reference that you want to explore. Try using the browse gallery button. As an activity I want you to try using any of the two or three options available in the **Popular** category. You will be amazed by the results that it will generate. You can refer to *Figure 2.16* to see the output.

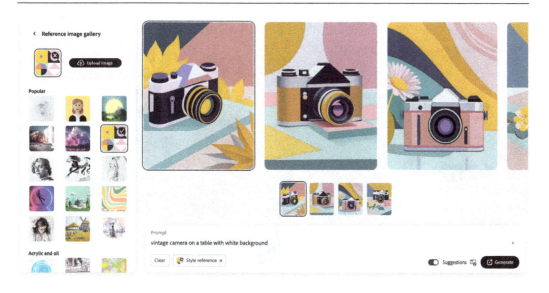

Figure 2.16 – Trying out other styles using Reference image gallery

Using your own images with Style Reference

You can also try to use your own image that Generative Match will try to generate. I tried one using a recent image I created also in Adobe Firefly with the following prompt:

```
8-bit orange cat with cool dark sunglasses.
```

Feel free to copy that prompt and apply it as your own.

Take note that it will analyze the image uploaded together with a notification that you own the image you are using with Generative Match. Using the same prompt, it surprisingly created this output as seen in *Figure 2.17*.

It has properly emulated the style indicated in the image, which is an 8-bit pixelate style; even the colors are copied accurately from the reference.

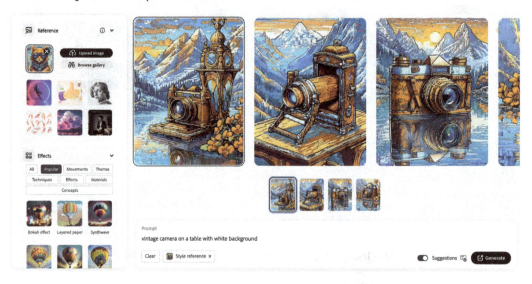

Figure 2.17 – Using our own reference image in Style Reference is like magic!

There you have it – using the Generative Match with Style Reference feature is truly magical; try it with other built-in samples as an activity for you to explore! I personally love the 3D style in the middle of the **Popular** style category. Next, we will learn about using effects.

Using effects in the Text to image module

If you decide that crafting effective prompts is not really your thing, you will be glad to know that Effects will help you a lot. Think of it as your comprehensive list of specific styles, in which you just need to click on one and it will apply it with your prompt.

As of the time of this writing, there are 124 effects that you can choose from, categorized into groupings such as **Popular**, **Movements**, **Themes**, **Techniques**, **Effects**, **Materials**, and **Concepts**. You can easily access all of them by clicking the **All** category in the filter buttons section.

This is one area where Adobe Firefly stands out – it has a user-friendly interface that will take out the frustration and let you focus on the thing that matters, which is creating. Once you apply one or more effects, it will be tagged below your prompt and a checkbox will be visible in the effects section noting that you applied it.

Using our basic prompt, we applied the **Cubism** (a style made up of blocks, which are fragmented and abstract), **Film Noir** (dark imagery), and **Pointillism** (small dots as a pattern to form an image) style. See *Figure 2.18* to see the generated results.

Figure 2.18 – Style tags are applied and a checkbox is in the Effects section

You can easily remove an applied style by clicking on the **x** button on the tag itself or clicking the **Clear styles** button to get rid of all of them in just one click.

In the next section, we will take it deeper by exploring more advanced settings that are harder to replicate easily just by using prompts.

Working with the advanced settings of Adobe Firefly

There are several advanced settings that you could try with Firefly. We will first discuss how to use color and tone effects.

Using color and tone effects

We have 10 options that you can choose for your images. You can use **None** as **default, Black and white, Cool tone, Golden, Monochromatic, Muted color, Pastel color, Toned image, Vibrant colors**, and **Warm tone**.

My favorites in the list would be **Vibrant colors**, **Pastel colors**, and **Monochromatic**. Unlike the **Effects** list, you can only use one at a time.

In *Figure 2.19*, you see an example of this in action.

Figure 2.19 – Using Vibrant colors as the Color and Tone option

In the following sections, I have cleared all the styles and settings and focused on each setting applied. Next, let's try changing the light effects and camera angle.

Changing lighting effects

Having the ability to change how the light interacts with your generated images is such a great option, as lighting has the power to convey a specific feeling visually through brightness, darkness, tone, mood, and atmosphere.

The ability to control light is as important as creating a statement, as photographers spend a great amount of time just learning this skill.

Try the following options in the list: **Backlighting**, **Dramatic light**, **Golden hour**, **Harsh light**, **Long time exposure**, **Low lighting**, **Multi exposure**, **Studio light**, and **Surreal light**.

My favorites in the list are **Golden hour**, **Dramatic light**, and **Low lighting**, which give stunning results. Try it yourself! Check out *Figure 2.20* for reference.

Figure 2.20 – Using Dramatic light as the Lighting option

Adjusting composition

Similar to how we can generate lighting effects, having the option to change how the composition works truly makes a lot of difference as you can visually adjust it using generative AI. You can choose **Close up**, **Knolling**, **Landscape photography**, **Macrophotography**, **Photography through window**, **Shallow depth of field**, **Shot from above**, **Shot from below**, **Surface detail**, or **Wide angle**.

My favorites are **Knolling**, **Shallow depth of field**, and **Macrophotography**. There are so many options now and that just makes me excited to explore more. See *Figure 2.21* to discover how this works.

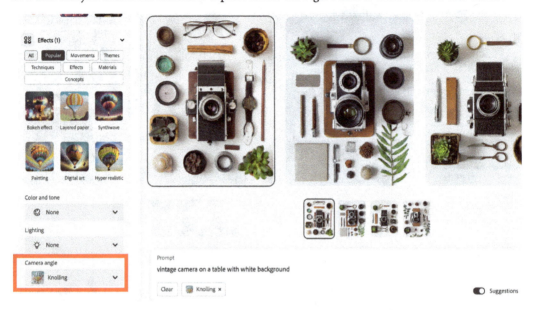

Figure 2.21 – Using Knolling as a Composition option

Advanced editing options within an image

Now that we have practiced mostly all of the available options, let's move our focus to the editing functions, which we can use within Adobe Firefly and Adobe Express.

With each image generated, you can access it by hovering your cursor over the top left section and clicking on the drop-down arrow, which offers the following choices (*Figure 2.22*).

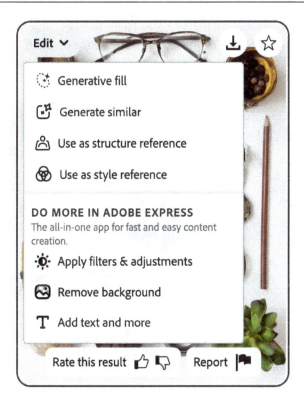

Figure 2.22 – Access to all editing options with Adobe-Firefly- and Express-related tasks

When you choose the options for Adobe Firefly, this will enable you to access the same image within the same window unlike when you chose those with the Adobe-Express-related options. Let's talk about using the following:

- **Generative fill**: This will allow you to do precise editing using brush tools to add, remove, or replace existing pixel elements within the image. We will be covering this in-depth in *Chapter 3*.

- **Generate similar**: Using this will trigger Adobe Firefly to create versions of the same elements being placed in the images which will closely resemble the original. Think of it as a way to refine and use it as an input to generate new versions based on the image as a reference. See *Figure 2.23* to gain visual clarity on this.

Figure 2.23 – Using Show similar enables you to create variations with the most accuracy

- **Use as structure reference** – enables you to use the selected image as a Structure Reference.
- **Use as style reference** – enables you to use the selected image as a style reference.

Extending your work into Adobe Express

The **Edit** option is dedicated to handing off your images directly to Adobe Express. It appears at the top left of the selected image. You can further enhance your image by applying filters and adjustments, removing backgrounds, and adding text.

Clicking any of the options in the section will guarantee another tab to open for Adobe Express to process. Check it out in *Figure 2.24*.

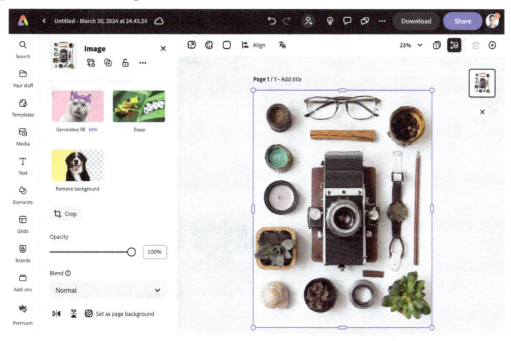

Figure 2.24 – Adobe Firefly handing the same image to Adobe Express for further editing

Using the Share Options

Back on the Firefly website, let's move into the **Share** button section and see what other functions we will get. If you don't know where it is located, it is on the top right of the user interface, next to the **Download** button. I have included *Figure 2.25* to guide you to its location.

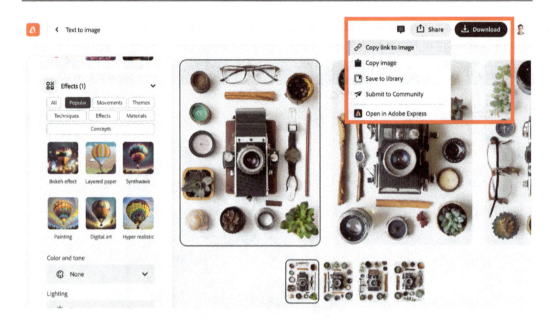

Figure 2.25 – Share options and other controls that you can use

Sharing your generated images via a link

If you want to share your generated images without downloading them, you can easily share them with anyone using the **Copy link to image** function – it will embed Content Credentials upon generating the link, and it will be directly copied into your clipboard.

The receiver of the shared link can view and also edit your creation further, including the prompt that you input with all of the options available at your fingertips as an Adobe Firefly user. I have included a shortened link to what I have created so that you can try it for yourself. Enjoy: `http://tinyurl.com/sharefireflylink`.

You can use the **Copy image** option to use it in another application, you cannot edit this further, and it will be a flat image as you use the **Paste** function, unlike the **Copy link** option.

Save to Library enables you to save the image in your Creative Cloud Libraries folder. This is only applicable to those who have an active paid subscription.

Saving your prompts as favorites

Whenever you want to revisit the input prompts at a later time, you can simply do so by clicking the hollow star icon at the top-right section of any image – this will save your work together with all the settings (use the filled star icon to remove from favorites).

You will be notified to view your favorites in a brief moment via a pop-up green button. See this in action in *Figure 2.26*.

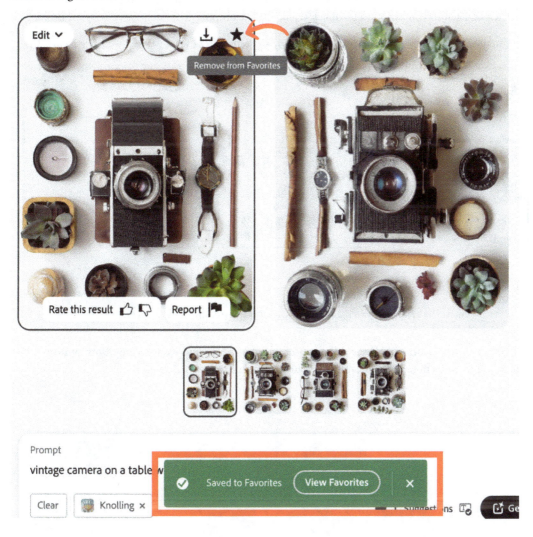

Figure 2.26 – Using the Favorites icon and View Favorites pop-up notification

If you missed it, you can click on the Adobe Firefly icon at the top left and click on **Favorites** (this is mentioned in *Chapter 1*).

Contributing to making Adobe Firefly better

In this section, we'll cover how to share your feedback (praises and concerns) with the Firefly team.

You will see buttons titled **Rate this result**, indicated by a thumbs up and thumbs down icon. You can click on each so you can send feedback regarding the outputs generated based on your prompt.

The thumbs up will enable you to provide the following responses of what worked well. See *Figure 2.27*.

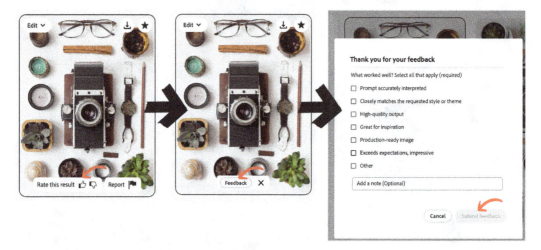

Figure 2.27 – Feedback process for the thumbs-up button

The thumbs-down icon will offer a different set of options for what went wrong. Both options will enable you to add a note when submitting your feedback. See *Figure 2.28*.

Figure 2.28 – Feedback process for the thumbs-down button

While there are safeguards on Adobe Firefly to prevent misuse or unintended results based on the prompts created, sometimes, even with good intentions, it may be misled into creating something that may violate specific guidelines within the Text to image module.

It is not perfect and often, some words may be flagged as false positives, given that most of it doesn't have enough context to decide on things even we humans can be susceptible to. You can send a report regarding a myriad of concerns by using the **Report** flag button.

Check *Figure 2.29* for the list of categories that are considered in violation of the following conditions:

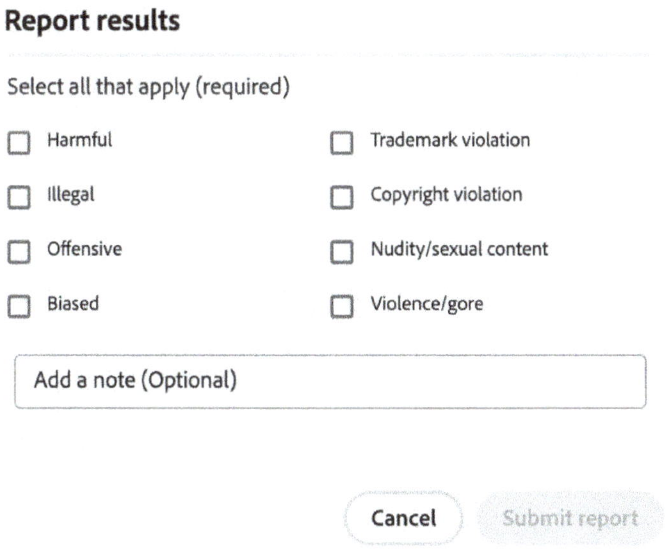

Report results

Select all that apply (required)

☐ Harmful   ☐ Trademark violation

☐ Illegal   ☐ Copyright violation

☐ Offensive   ☐ Nudity/sexual content

☐ Biased   ☐ Violence/gore

Add a note (Optional)

Cancel   Submit report

Figure 2.29 – Options that you can provide to report specific concerns when generating images

Now it is your turn to contribute to keeping Firefly safe by reporting results that may cause harm to others. You can also provide feedback on the assets it has generated making it better as it improves learning. You can also opt out if you want. Ultimately, the choice is within your own decision.

Summary

There you have it – we have covered a lot of ground in this chapter, from crafting your own prompts, using the Text to image module to generate outputs, using Generative Match, Structure Reference, and Style Reference with Reference image gallery, adjusting the color, tone, lighting, and compositions, exploring your options to download, share, and edit in Adobe Express, to providing feedback to help the Firefly team offer better results and maintain a safe environment. In the next chapter, we'll learn how we can use the Generative fill module.

3

Editing Images Faster using Generative Fill

In this chapter, we will explore how we can edit images faster using **Generative Fill**. We will be generating images on top of existing pixels, adding objects with precision, removing unnecessary ones, changing brush settings such as size, hardness, and opacity, and downloading the finished image as output.

In this chapter, we will discuss the following topics for the Adobe Firefly web app:

- Using Generative Fill
- Changing brush settings
- Changing the image background
- Downloading and sharing your image

Using Generative Fill

Way back in the day, it was a hard task for designers to edit images easily. They had to maintain aspects such as color blending, shadows, lighting, style, and other factors to make the perfect convincing output.

With Generative Fill technology in Adobe Firefly, this has become a thing of the past and happens without you breaking a sweat or needing a cup of coffee, because you need to wait for it to process, enabling you, the creative, to achieve realistic results. With the commercial release, you can ensure that you can push creative boundaries.

Think of Generative Fill as a magic tool for all of your images – while it is still in its infancy for the most part, it will greatly improve as the months or years pass. It is just a matter of time until we find out how good it is going to get.

Benefits of using Generative Fill

Let's look at a list of some of the significant benefits that Generative Fill can offer:

- Without Generative Fill, you would need to access stock image websites and search for hours upon hours to find that perfect asset, and then copy it over (individually) to the existing image you are producing; and you would have to do this almost every time the image asset does not match or blend in to create that convincing mix.

- Using Generative Fill, you only need to make a brush selection and supply specific prompts to add the elements needed in your design without acquiring technical knowledge in making your design work together visually.

- It frees up a lot of time because the AI will blend it automatically and generate the needed images as supplied in the prompt. No more juggling layers and switching panels as seen in the left portion of *Figure 3.1* – the right side is exclusively using Generative Fill.

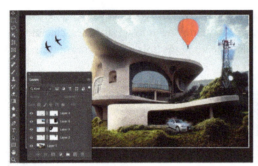

Figure 3.1 – Comparison in compositing images without
Generative Fill (left) and with Generative Fill (right)

What can you do with Generative Fill?

There are a lot of creative options that you could experiment with using Generative Fill:

1. **Insert and remove objects seamlessly**: Select an area and it will automatically insert the requested prompt into the image without the need to access lots of adjustment controls.

2. **Generate background images**: It is now easy to change any background behind your subject, and then generate a new one based on a text prompt.

3. **Extend your canvas**: Regardless of the target size of your deliverable output, you can make it fit in just a few clicks (this is not available in the Firefly web app).

4. **Have fun and explore your creative options**: With Generative Fill, you can have the freedom to change almost everything, from clothes to accessories.

Now, let's explore how we can use Generative Fill in the Firefly web app!

Accessing the Generative Fill module

You can use Generative Fill in Adobe Firefly by accessing the web service at this website address: `https://firefly.adobe.com/`. You need to sign in using your Adobe ID. Now, you can follow these steps to use this module:

1. Click on the **Generate** button in the **Generative fill** section, as seen in *Figure 3.2*.

Figure 3.2 – The Generative fill section with the Generate button

2. It will open the **Generative fill** workspace in which you can choose from any of the sample images available, or upload an image that you own for you to edit.

    As an activity, we will click and choose the lighthouse image in the middle so we can get started learning. See *Figure 3.3* for your guidance.

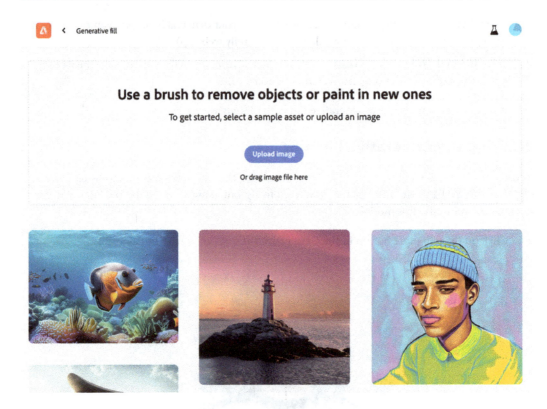

Figure 3.3 – The Generative fill module with sample images you can use

> **Tip**
>
> A JPG file format can provide easy editing tasks – I have tested uploading a 30 MB JPG (13583 × 5417) file and used Generative Fill and it takes more time to process due to file size together with the speed of your internet connection. Generative Fill in Adobe Firefly accepts only JPG, PNG, and WebP file formats.
>
> You can also upload a PNG file format that offers transparency and higher-quality output if you are into producing print materials. WebP is a modern image format that provides superior lossless compression for images on the web but is not as popular as other options at the time of this writing.

3. Clicking it will enable you to access the **Generative fill** interface. You can check the floating controls on the left – this will enable you to carry out operations such as inserting and removing objects represented by the brush tool.

    It also has the **Pan** tool, so you can reposition your image for precise editing. By default, the **Insert** operation is active. See *Figure 3.4* to familiarize yourself with the interface.

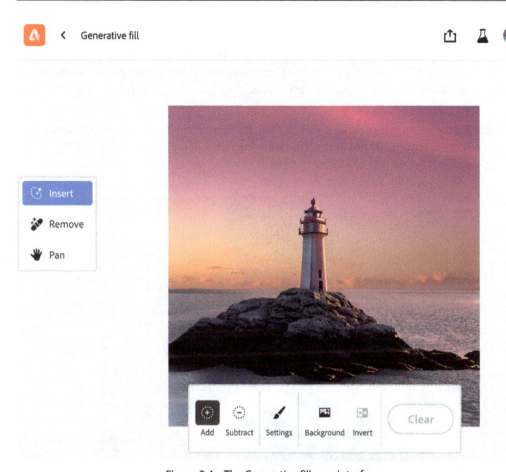

Figure 3.4 – The Generative fill user interface

The bottom interface offers controls to add, subtract, adjust your brush settings, remove the background, invert any selection, and clear (cancel) any active pending operation.

Next, we will try and insert objects to make this image more interesting.

Adding and removing objects with Generative Fill

The little details make the difference, and adding them can take minutes. Unlike other image editing techniques, Generative Fill allows you to do the necessary editing within the application itself and not jump in from another such as an internet browser or a local system file.

It is designed to let you focus on imagining and making the necessary adjustments without interrupting your creative flow. Just select an area using the brush, type the object you want to add using prompts, and click the **Keep** button.

Let's get into the detailed steps on how to do this now:

1. Upon opening any sample image by default, you are in the **Insert** mode. When you move your cursor in the canvas, you will see that it indicates a big circle mimicking a brush and the **Add** button is active on the bottom part of the interface. So, most of the work is already established on your end.

2. Brush the top of the image, making sure that you have covered as much as the needed portion of the sky in the image. Do not worry if the following image does not precisely match the result you see. Check out *Figure 3.5* for your reference.

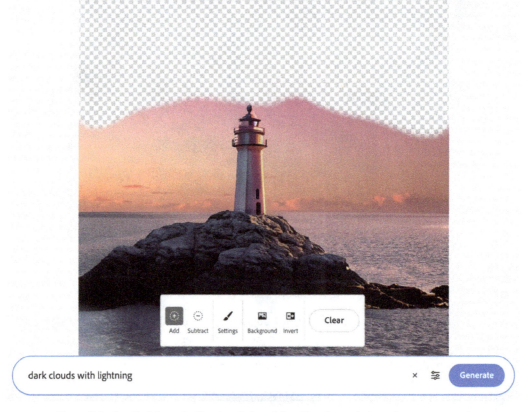

Figure 3.5 – Applied the selection brush for adding the elements based on our prompt

3. After selecting the portion of the image you want to select, a textbox will appear for you to type in the prompt you need – in our case, I have supplied the needed input:

```
dark clouds with lightning
```

4. Click the **Generate** button. After clicking the **Generate** button, it will give you three options of resulting images that you can choose from. You can select one of them and it will preview the result in the canvas for you to decide. If you are not satisfied with the results, you can easily click the **More** button to generate more possible renditions of the requested prompt, as seen in *Figure 3.6*.

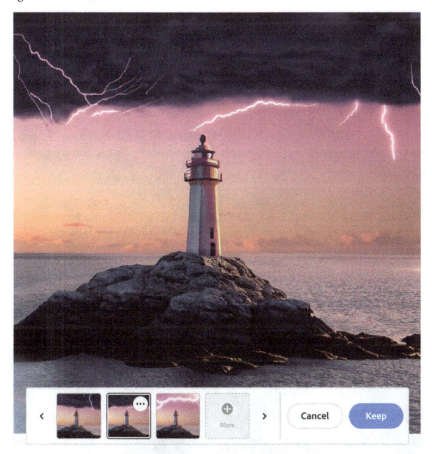

Figure 3.6 – Result of inserting objects based on prompts

5. Choose the one that you like and click the **Keep** button or **Cancel** if you want to start over again. This is best for situations where your selection with the brush needs more adjustment to ensure that the right areas will have your new image applied to them.

By using this process, you will understand that it can open a lot of possibilities in doing image editing, especially since it has done a good job of incorporating the needed objects.

> **Note**
>
> You can use the [keyboard shortcut to increase brush size and] to decrease it; zoom into the areas you need to brush to make a more precise selection.
>
> You can also use any device-specific gesture for you to navigate within the image. On my end, I am using my trackpad – pinch and swipe functions offer me a faster workflow.

For you to practice the activity, I added two more elements to the image, repeating the process with each prompt. These two of them are as follows:

```
Sailboats
birds flying
```

Feel free to place them in your canvas and refer to the suggestion in *Figure 3.7*.

Figure 3.7 – Added two more elements using the Insert function

Next, we will learn how to tweak the settings.

Changing the Generate Settings

When you input any prompt in the **Generate** text box, you can access the settings by clicking the settings button (to the left of the **Generate** button), which will enable you to access the following functions:

- **Match shape**: The default is **Freeform**, which will create a loose matching of generated output, or you can use **Conform**, so it will produce a more accurate result based on your brush selection.

- **Preserve content**: The default is **New**, which will provide very different content to your original selection. Whenever you want to retain the original, you can adjust the slider to the right.

- **Guidance strength**: This refers to how closely your result will be generated from an original asset or working strictly with the prompt as a priority.

You can check these out by referring to *Figure 3.8* after doing the activity.

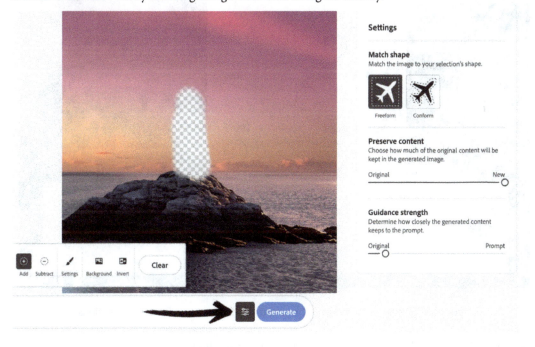

Figure 3.8 – Generate Settings

Let's continue and see how can you customize your brush to make your control accurate for precise selections.

Changing brush settings

There will be times when you will want to control the size, hardness, and opacity of your brush, and this is done using the **Settings** button in the interface. This is good for when you want to make a better application of Generative Fill in your images.

If you want to adjust to a custom brush preference, you can click the **Settings** button on the lower part of the interface – this will enable you to adjust the following parameters.

As you can see in *Figure 3.9*, there are several brush settings:

- **Brush size**: This can help you define how big or small the affected region is.
- **Brush hardness**: This is how sharp the edge of the brush is upon application (**feathering**).
- **Brush opacity**: This is how much transparency or opacity is being used.

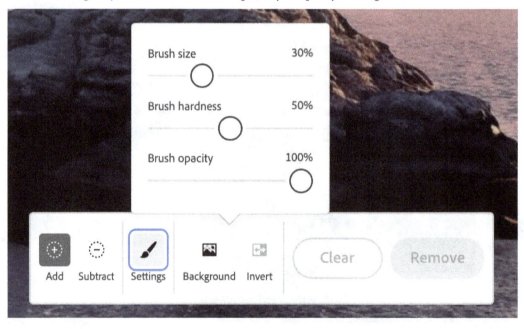

Figure 3.9 – The available brush settings you can choose to make better selections

Now, how do we pick the right combination of settings? Let's discuss this next.

Making better selections

As we learn how to customize the settings of our brush selection, it is important to know that using a combination of size, hardness, and opacity can benefit the use of Generative Fill in creating better images. Here are some tips you need to consider to make this possible:

1. Carefully select using the **Size** property, overextending a few areas with your brush to make sure that you cover everything you need to change. See *Figure 3.10*.

Figure 3.10 – Adjusting the Brush size setting

2. To get better blending results, try experimenting using the **Hardness** property. It makes your brush respond to lighter pressure, giving you control of specific parts of your selection. See *Figure 3.11*.

Figure 3.11 – Modifying the Hardness level of the brush

3. If you want to preserve most of your design assets generated and wish to only create versions of the same kind, you can use the **Opacity** setting. Just brush the needed part with a low setting and click the **Generate** button without typing any prompt, as seen in *Figure 3.12*.

Figure 3.12 – Controlling Opacity enables you to create very
subtle revisions and maintains most of the area

With these options, you can craft more advanced selections and results to make Generative Fill work to your advantage, especially if you want to create more realistic outputs.

Next, let's try to remove objects.

Removing objects seamlessly

As simple as inserting objects, you can also remove objects using the **Remove** mode in Generative Fill. Let's go back to our initial image and try to remove one of the sailboats that we have inserted. Make sure that you are in the **Remove** function on the left floating option bar to properly do this.

Brush the needed elements you want to remove – in our case, you will see that I have selected additional elements (*Figure 3.13*).

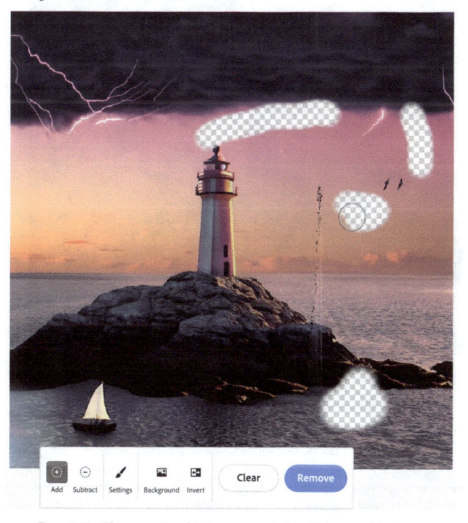

Figure 3.13 – When removing objects, you can select multiple areas of the image

Upon clicking the **Remove** button at the bottom of the interface, you will notice that you do not need to supply any prompt and it still generates three options that you can choose from. When you have finally chosen the needed result, click the **Keep** button to confirm your action. You will see the intended output in *Figure 3.14.*

Figure 3.14 – The final output for our initial image exercise

Now that we have successfully learned how to remove objects, let's try changing backgrounds to transport us to other worlds. Let your imagination run wild and, most importantly, have fun!

Modifying the image background

Changing the background can be a tedious task because it will require you to know how to use selection tools offered in most image editing software. One of the most common uses of this workflow is when we composite images to create assets that have never been together in the first place.

Changing the background

Using the **Select** subject in Photoshop enables you to make an automatic selection using AI, offering faster workflow, and Firefly takes it further by simplifying it with just a few clicks. Here are detailed step-by-step instructions on how to do this:

1. In this example, we need to go back and click on the top-left portion in which the Adobe Firefly icon is located, and access the Generative Fill feature again by clicking the **Generate** button.

2. Let's look for the image of a woman wearing an orange jacket. Click on it to launch the **Generative fill** module. See *Figure 3.15*.

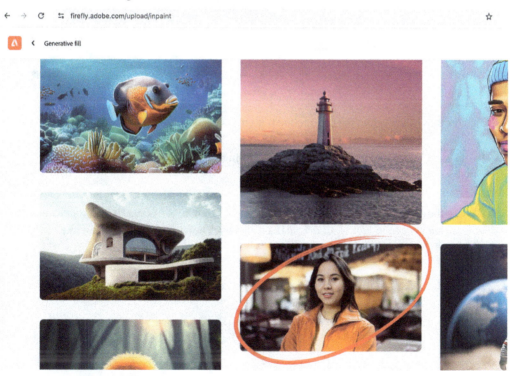

Figure 3.15 – Selecting the woman in the orange jacket

3. Click on the **Background** button and it will automatically detect the subject and select the background for you. It may take a moment because it has to analyze the image. When done, it will look something similar to *Figure 3.16*.

Figure 3.16 – By pressing the Background button, it will select the subject of your image

4. In our case, let's try to generate a background using this prompt:

    ```
    busy futuristic library with students
    ```

5. Choose the appropriate image background that suits your style because it will not produce the same result as seen in *Figure 3.17*. This is a subjective choice that you have to make.

Figure 3.17 – Changing the background based on our prompt

6. Click the **Keep** button when satisfied with the result, or click the **Cancel** button if you want to start over and type your own prompt.

It is not just the background that you can easily replace – you can also replace the subject matter in the image by using the **Invert** action. Let's try it in the next section.

Selecting the subject using the Invert option

We will be using this as a technique to *replace the subject* rather than the background. Common use cases of this are being used when we want to provide alternate options with the subject as the target we want to replace.

Let me show you this in action.

1. With the same image, click the **Background** button and then click the **Invert** selection button. It will select the subject, thus enabling us to replace our subject faster. Using this technique will save you time rather than brushing it all again.

2. Type the following prompt:

    ```
    woman in red jacket and baseball cap
    ```

Do not hesitate to click the **More** button to generate more options for this until you are satisfied with the result. You can see this in action with our final output in *Figure 3.18*.

Figure 3.18 – Generating a new subject using the Invert function

You can also do the same procedure on your own photo. Click the **Upload image** button or drag the image onto that portion.

You can brush parts of the image where you may see some inconsistencies and make it to correct common mistakes that it has made. You will see some extra pixels in the outline of the newly generated image, but you can use the **Remove** function and instantly brush them out.

Next up, let's learn how we can use the images we generated.

Downloading and sharing your image

In this **Generative fill** module, when you are done with all the necessary steps to finalize the image, you can click the **Download** button at the top-right section of the interface. The process also takes a similar approach when downloading the self-contained file in the Text to image module.

You will be prompted by the **content credentials** window and you need to click the **Continue** button for the download process to be executed.

A new file will be generated with the .png extension that you can use as a self-contained format. This is the only format available for download as of this time of writing.

Sending your image to Adobe Express

Similar to the download process, the controls to send your image to Adobe Express are located in the same location, indicated by a triple dot button. Here, you can pick the following options: **Apply filters & adjustments**, **Remove background**, and **Add text and more**. You can access the following options as seen in *Figure 3.19*.

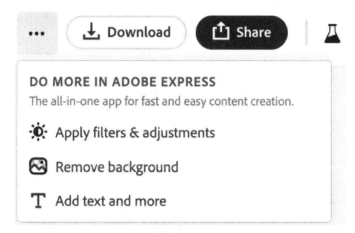

Figure 3.19 – All the available options for Adobe Express using the Generative fill module

Take note that it will open another tab in your browser by doing this operation.

Sharing options within Generative Fill

When you click the **Share** button, it will provide you access to hand it over to Adobe Express. The only difference is that it will not open any specific controls such as **Adjustments**, **Text options**, or **Remove background**. It will open the image with all the possibilities you can use inside Adobe Express. See how you can access this in *Figure 3.20*.

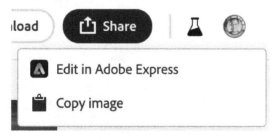

Figure 3.20 – Options for the Share button

Last in our options is the ability to copy the image into our clipboard, which will enable you to repurpose the image into other applications, rather than downloading it and importing it. This is a time saver for those who are into working fast with the images they have created.

> **Tip**
>
> Using Generative Fill in the web app will enable you to understand how it works with the current integration of the same feature in Adobe Photoshop. After downloading the final image file, you can import it to your existing design canvas as a .png file.

That's it – all of the possible buttons and options you can do with Generative Fill inside the Adobe Firefly web app. I highly encourage you to try it even once so you can experience and understand how it works.

Summary

Editing images nowadays can be done in just a matter of seconds due to the help of generative AI, and it is still improving with every released version.

In this chapter, we have gained first-hand experience of how effortlessly you can insert or remove elements in an image using the brush tool, edit the settings of your brush for precise selection, completely change your background images, download images as a standalone file, and repurpose and hand your images over to Adobe Express for further editing.

I hope you had a great time learning about the Generative Fill feature inside of Adobe Firefly web app.

In the next chapter, we will create stylized text effects! It is going to be exciting!

4

Creating Stylistic Text Effects in Adobe Express

In this chapter, we'll delve into the art of crafting and personalizing stylistic text effects. Together, we will explore how we can use text effects in our day-to-day design workflow and review some other possible case studies that could benefit us in generating unique designs that stand out.

Text effects can be readily integrated into your projects and are available in Adobe Express, which offers fine-tuning and more editing features.

By the end of this chapter, you'll not only master the intricacies of text effects but also possess the skills to produce striking visuals that can transition into the next phase of your creative process.

In this chapter, we'll cover the following topics:

- Understanding how text effects work and some practical use cases
- Learn how to get into the advanced settings and customize each text effect's output

Using text effects

Imagine creating decorative, stylized, and beautiful display text in seconds by incorporating text prompts, saving you time while having a new set of creative tools at your disposal. Well, you don't need to imagine with **text effects**.

Here are some of the use cases of text effects:

- Make unique and attractive design elements for posters, flyers, logos, websites, social media pictures, and more
- Develop customized and unique-looking personalized message fonts, greeting cards, and movie text titles
- Create interesting design elements for learning materials such as books and worksheets

All it takes is your imagination, together with a specific prompt describing what you want your text to look like. Let's learn some recommendations that you need to know to make text effects much more effective.

Recommendations for using text effects

Here are some best practices for applying text effects to enhance a design layout:

- Limit the number of characters to 10 or less for greater design impact
- Use text effects in minimalistic design layouts that will make text effects stand out as a design element
- Useful for conveying information by using decorative art or Display Type that catches the attention of the viewer
- Use text effects with large pieces of text, which will give emphasis to the textures generated
- Text effects are best used when working with a specific design theme

You can get an idea of these best-practice recommendations in action by checking *Figure 4.1*.

Figure 4.1 – Example uses of text effects in designs

In the next section, let's create our own text effects to practice what we've learned.

Creating text effects in Adobe Express

For your creative projects, you may want to add more inviting and attention-grabbing elements to your designs, especially if you are working with layouts intended for social media campaigns. What better way than to use text effects to accomplish that goal?

You can use text effects exclusively in Adobe Express at `https://new.express.adobe.com/`. You need to sign in using your *Adobe ID*, and then you can get started:

1. Let's create a new text effect from scratch. Scroll down a bit on the website and click on the **Text effects** section, as seen in *Figure 4.2*.

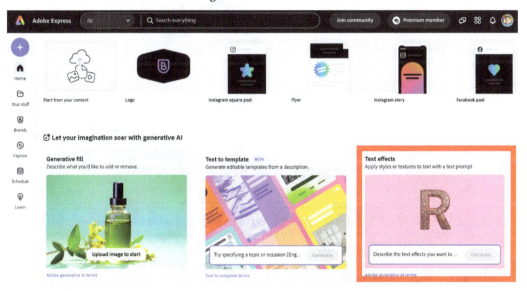

Figure 4.2 – Accessing text effects on the Adobe Express home page

2. A new browser tab will open together with a blank Adobe Express project. It will automatically display the results of an example prompt, as seen in *Figure 4.3*

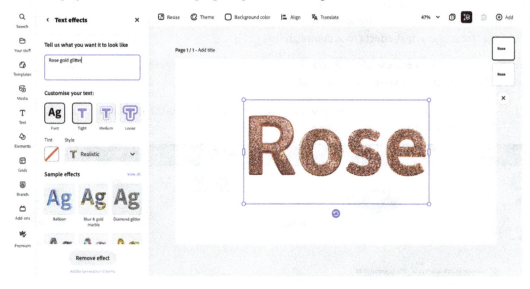

Figure 4.3 – A new blank project with a generated text effect

Now that we learned how to access text effects in Adobe Express from scratch, let's try adding another one by using another method. This is beneficial for you to know if you want to add text effects to your existing Adobe Express projects.

Adding another text effect

There is another way that you can add text effects to your Adobe Express project. This is best when you have an already existing design and want to add more changes. You can easily add text effects by using the following method:

1. Click on the **Text** button on the left side of the Adobe Express interface.
2. The **Text** panel on the left side will open with a **Text effects** button available on the bottom portion of the interface, as seen in *Figure 4.4.*

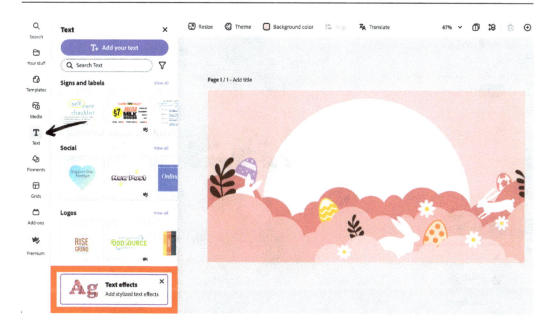

Figure 4.4 – Creating text effects using the Text button in Adobe Express

3. Click the **Text effects** button and it will offer you a list of example text-effect prompts that you can start with. Click any one of the examples to edit its options.

4. Input the text in the canvas and modify it as needed.

> **Important tip**
>
> If you accidentally closed the text effects panel by clicking on other design elements, you can easily go back by clicking the **Text** button, scrolling down the **Text** panel (the left panel), and clicking on the **Text effects** button, as seen in *Figure 4.5*.

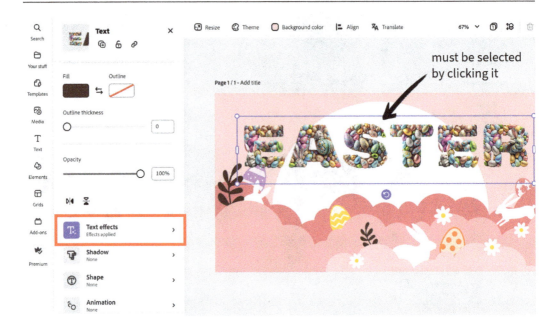

Figure 4.5 – Accessing the text effect's configuration options

You are now equipped with the necessary knowledge to access text effects in Adobe Express. Let's next try to make unique creations by customizing our text effects to suit our needs.

Customizing your text effects

When the results generated do not match what you have in mind, you can try customizing them using the left side of the interface (**control panel**). Let's explore and try it ourselves, starting with changing the text effect's input and prompt.

Changing the text effect's input and prompt

1. Let's change the text effect's input by clicking on the text on the canvas and changing the word to Leaf.

2. Click the textbox on top of the control panel and type in the following prompt:

    ```
    Fiddle leaf fig
    ```

3. Click the **Generate** button to see the results, as shown in *Figure 4.6*.

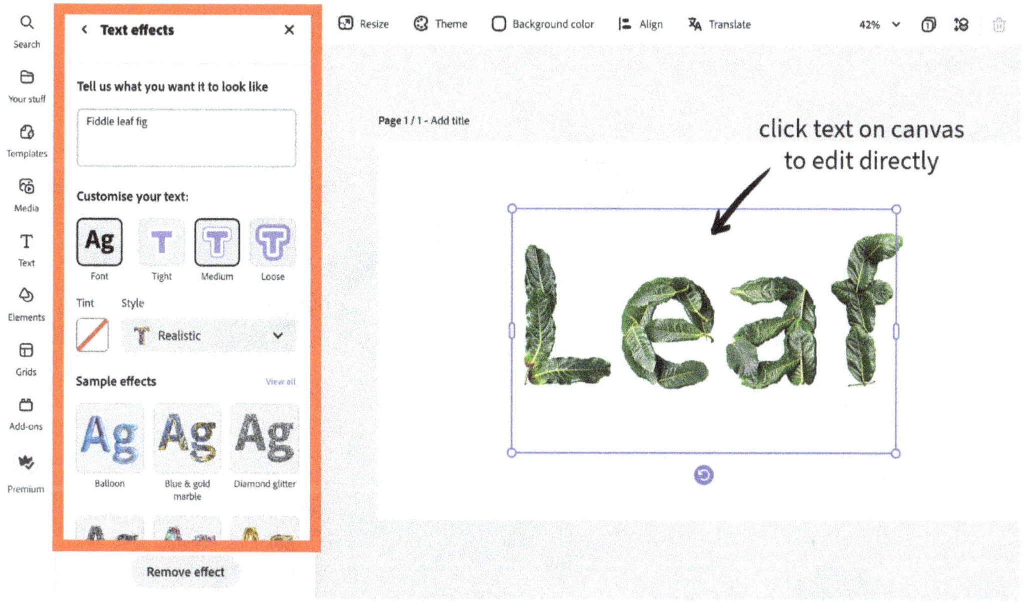

Figure 4.6 – Text effects controls inside Adobe Express

> **Tip**
>
> Click the text on the canvas itself and not the prompt input box.
>
> This will ensure that you are changing the text itself, and not the text effect prompt.

Changing the font

Just as important as choosing the right prompt is choosing the right font. Fonts enable you to create a placeholder for your text effects to be placed on. While Adobe has given us lots of built-in choices, in my opinion, it is a bit limited for now. You cannot add any fonts of your own at the moment.

You can explore more of the fonts available with the following steps:

1. Click the **Font** button.

2. As an activity, let's try using the **Cooper Black** font. You don't need to click on the **Generate** button, as it will update the font automatically. *Figure 4.7* illustrates these actions.

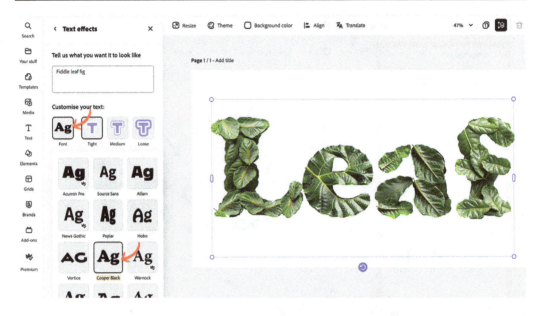

Figure 4.7 – Using the Cooper Black font as we change our type style

I would recommend that you choose a font style such as **Source Sans**, **Alfarn 2**, or any bulky type of font to give emphasis on the overlaid texture of the text effect.

Choosing a thin font can only be recommended when you are using it against a seamless or flat color background with minimal elements that complement the text, or when the focus is the text itself. You can see this in action in *Figure 4.8*.

Figure 4.8 – Examples of visual impact when changing font styles

That's how easy it is to change the font in your text effects. Next, let's try another way to customize your text effects with the Match shape option.

Changing the Match shape option

Match shape allows you to create expressive and decorative designs with your text. It is best used when you have fewer than 10 characters because it may become visually overcrowded otherwise.

Utilize the **Match shape** feature to determine the snugness with which the design conforms to the characters' outlines. Opting for a looser text effect will enable it to extend beyond the usual character boundaries. In *Figure 4.9* you can see the contrast between both options.

Choosing the right type of font is also an important consideration. When you use Serif fonts, it tends to generate a looser output than San Serif fronts. It is up to your subjective taste to determine which will work most cohesively with your overall design.

Figure 4.9 – Using the Loose option (top) in comparison with the Tight option (bottom)

Using the **Tight**, **Medium**, and **Loose** options provided in the Match shape controls, you can extend the generated effect past the font's borders. Each generated output is unique, so experimentation is the key to creating a good text effect.

Using text styles

So far, you've seen how to make unique text effects using prompts describing how the text should look. Using **text styles** provides you with additional controls to take your customization further. Let's examine each of the built-in styles on offer. You can try them on your own as an activity to practice.

Embellished enables you to create text effects that add elements of detail and extend certain features while having a defined structure underneath. It is an expressive form. See *Figure 4.10* to examine it closely.

Figure 4.10 – Using the Embellished text style option

Realistic enables you to create text effects that simulate how elements in the real world appear in the best way possible. See *Figure 4.11* to examine it closely.

Figure 4.11 – Using the Realistic text style option

Pencil Sketch enables you to create text effects that simulate a pencil sketch style without the use of color. See *Figure 4.12* to examine it closely.

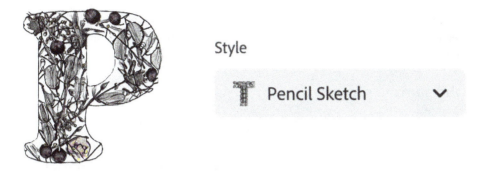

Figure 4.12 – Using the Pencil Sketch text style option

Neon enables you to create text effects that simulate a neon glow as an overlay. You can use the **Tint** option to explicitly indicate the color you need. See *Figure 4.13* to examine it closely.

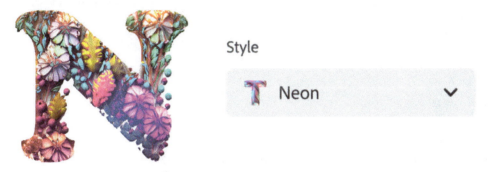

Figure 4.13 – Using the Neon style option

Colorful fineliner enables you to achieve a style similar to the Pencil Sketch style, but with colors included. See *Figure 4.14* to examine it closely.

Figure 4.14 – Using the Colorful fineliner style

This section covered a lot of special features to further customize your text effects. But you can also explore even more without needing to get too technical by using the built-in sample effects! Read on to find out more.

Using the sample effects

When you want to apply text effects without typing any prompt, you can use the built-in sample effects in Adobe Express. So what are sample effects?

The **Sample effects** panel offers you some premade text effects with the prompts included. You can click the **View all** button and you will be shown categories including **Reflective**, **Natural, Abstract**, **Animals, Textures, Arts and crafts**, **Food, Materials, Distressed, Digital, Painting, Flowers**, and **Fabric**.

You can see the comprehensive list of effects all together on one page using our cheat sheet in *Chapter 8*.

To select an effect, simply click on any of the sample thumbnails. In our case, we chose **Gold drip** and clicked the **Generate** button. Take a look at *Figure 4.15* to see this in action.

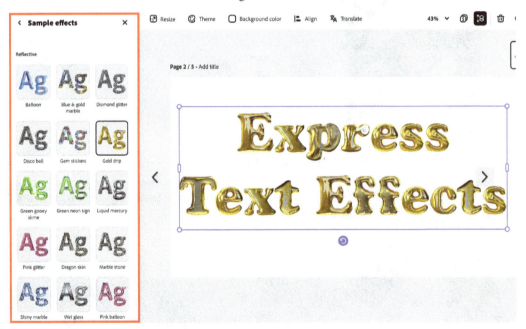

Figure 4.15 – Using the Gold drip built-in sample prompt

Sample prompts are best used as a starting point to describe specific elements you want to incorporate as textures in your text. The samples allow you to save time and act as a guide, together with the categories, to discover possibilities that you can input using a prompt.

Samples also function as a prompt-learning feature for beginners who are not well-versed in describing and generating textures based explicitly on words. One of the use cases I often employ sample prompts in is generating nature textures, which offers great results that can be added to Photoshop or Express designs.

> **Important tip**
>
> You can also use the **Text layout** options in Adobe Express to make amazing text effects. You can choose from **Dynamic**, **Circle**, **Arc**, and **Bow**.

Oh! I almost forgot… How do you export text effects and import them into your design tools, such as Adobe Photoshop or Illustrator? That is coming next.

Exporting text effects

As of now, there is no simple or quick way to export text effects to other design applications. One way that I have tried is the following:

1. In Adobe Express, click the **Background color** button, and set it to **No fill**. See *Figure 4.16*.

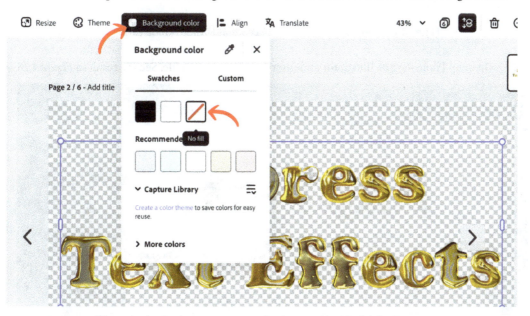

Figure 4.16 – Setting a transparent background inside Adobe Express

2. Next, go to the **Download** button and choose **Transparent PNG**. See *Figure 4.17*.

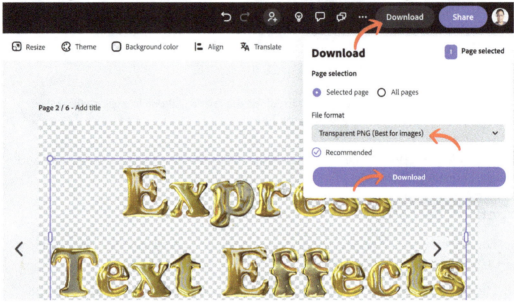

Figure 4.17 – Setting a transparent background inside Adobe Express

3. Open up Photoshop or Illustrator and insert the downloaded file. See the result in *Figure 4.18*

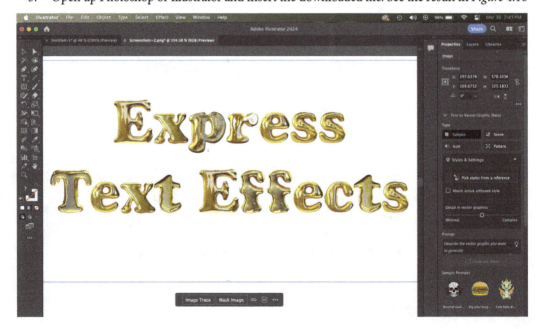

Figure 4.18 – Adding a file with a transparent background

There you have it! I hope you are excited to explore and apply text effects in your upcoming projects! I will see you in the next chapter as we use Generative Recolor in Adobe Illustrator.

Summary

With all of the information discussed in this chapter, you can now easily create great-looking text effects to add creativity when you design your next project.

We covered how to create text effects and how to customize them using options such as sample prompts and the Match shape option. We also saw how to select fonts, styles, and colors to use while editing in Adobe Express and how to export text to Photoshop and Illustrator.

In the next chapter, we will explore how colors can be created with prompts.

5

Exploring Color Options with Generative Recolor

In this chapter, we'll explore how we can create color variations using personalized prompts. We will cover the process of generating custom color schemes, use built-in sample prompts, examine color harmonies, and download our final output.

You will learn how easy it is to create and ideate colors using the Generative Recolor feature exclusively available in Adobe Illustrator.

We have included simple hands-on learning exercises to help you better understand how color manipulation works. Whether you are new to this field or have some prior experience, this chapter equips you with the basics to get you up and running with flying colors!

In this chapter, we'll cover the following topics:

- Creating color variations using custom prompts
- Exploring ready-made sample prompts
- Aligning color treatment by using color harmony rules

Technical requirements

When using the desktop application of Adobe Illustrator, you have to take into account the necessary processing power required to get smooth operations in your workflow. Use the following link to check whether your system can provide optimal performance: `https://helpx.adobe.com/illustrator/system-requirements.html`.

Why use Generative Recolor?

Color plays a big part in making effective designs, and simulating colors is a big deal for designers to explore possibilities, making it a great part of the workflow to consider.

With Generative Recolor, you can easily preview color variations for your product packaging, posters, or artwork. It can be a time-saver in the following tasks:

- Achieving a retro or vintage esthetic for your images by altering colors to sepia, black and white, or other nostalgic color schemes

- Crafting a mood board by harmonizing colors across various images to convey a specific theme or mood

- Enhancing accessibility by modifying colors to accentuate differences, making your image more perceptible and user-friendly

The following figure is an example of color variants available in this case:

Figure 5.1 – Color variants for deliverables such as certificates, illustrations, logos, and so on

Using a traditional approach of selecting each vector path in your designs, commonly tens to hundreds of layers, can be time-consuming as each change needs you to open panels and controls that may introduce consistencies.

As shown in *Figure 5.2*, with Adobe Illustrators' Recolor feature, you need to select multiple paths (indicated in blue), open the **Recolor** panel, drag sliders, and add colors, all of which take a lot of time and effort on the part of the user.

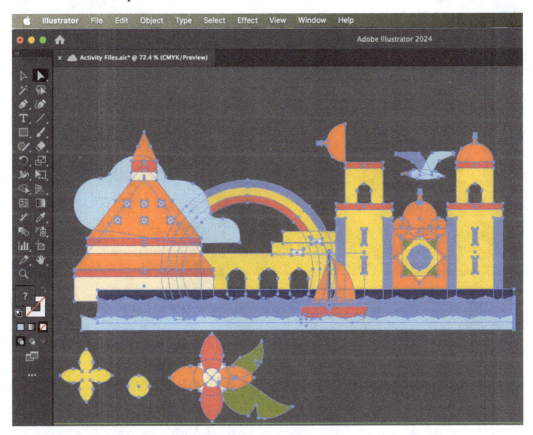

Figure 5.2 – The tedious task of creating color variations in Adobe Illustrator

In Adobe Illustrator, the **Generative Recolor** panel has also been introduced as shown in *Figure 5.3*, as a separate panel.

Notice that clicking one of the variations will immediately take effect in your selected design without the need to manually apply the changes one by one – a total time-saver and creativity booster!

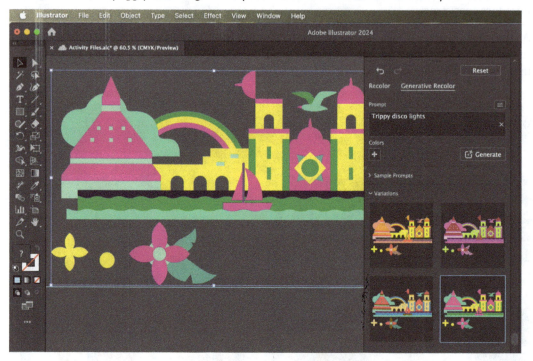

Figure 5.3 – The Generative Recolor feature inside Adobe Illustrator

Next, we will explore Generative Recolor exclusively in Adobe Illustrator to get us started and understand it more. Let's take a look at the following steps to access Generative Recolor.

Accessing Generative Recolor in Adobe Illustrator

Generative Recolor enables you to use Generative AI from Adobe Firefly, simply typing in a prompt that is applied to your vector drawings. This feature has been added only recently to Adobe Illustrator at the time of writing and is so easy to learn. The following steps will get you started:

1. To save time, you can download and open the Illustrator file I used in this chapter using this link: `https://tinyurl.com/generativerecolorfile`.

2. Select the drawing paths needed by pressing *Cmd/Ctrl + A* to select them all.

3. Go to the **Edit** menu, hover your cursor over **Edit Colors**, and click **Generative Recolor**, as shown in *Figure 5.4.*

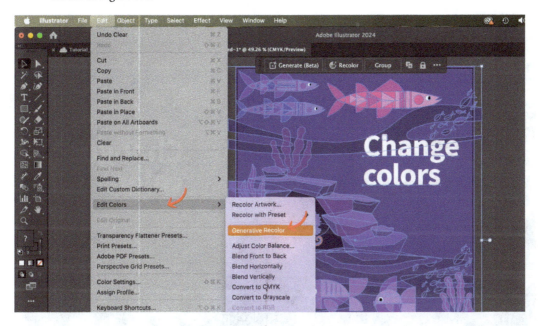

Figure 5.4 – Accessing Generative Recolor inside Adobe Illustrator

4. Click on any of the sample prompts or type in your own custom prompt to apply it to your current project. See *Figure 5.5* for the final result.

Figure 5.5 – Applying Generative Recolor inside Adobe Illustrator

> **Tip**
> When writing your prompt in Generative Recolor, use clear and simple words. Keep it short, around 2-4 words, to be most effective.

Try it yourself. As an activity, I have supplied some prompts you can use to explore:

```
calming forest green
tranquil underwater sea shade
soft pastel dream
```

> **Note**
> The result of Generative Recolor is only temporary, so make sure to select and *apply* the required color changes.

How did it go? Have you tried all of the sample prompts provided? You can now use the built-in sample prompts, which we will cover in the next section.

Changing colors using the sample prompts

By default, there are nine sample prompts to give you ideas on writing your own prompt. As an activity, feel free to click each of the samples and it will automatically change the content of the textbox in the **Prompt** portion of the interface. To change the output of the asset itself, you need to do the following steps:

1. Navigate to the **Sample prompts** panel of the **Generative Recolor** interface and select a prompt. In our case, let's click **Salmon Sushi**. It automatically creates color variations as the sample prompt is clicked.

2. Check the variations displayed, and choose one that suits what you're looking for. See the recommended result in *Figure 5.6*.

Figure 5.6 – List of sample prompts in Generative Recolor

That's it! You now know how to use the sample prompts in Generative Recolor within Adobe Illustrator. To get more insight on how to appreciate colors, let's explore the color harmony rules outlined in the next section.

Using color harmony

Color harmony means picking colors from the color wheel that follow specific rules, such as being complementary or analogous to each other. This is where color theory plays a big role in helping us understand how to generate certain combinations.

You can check *Figure 5.7* to review this concept:

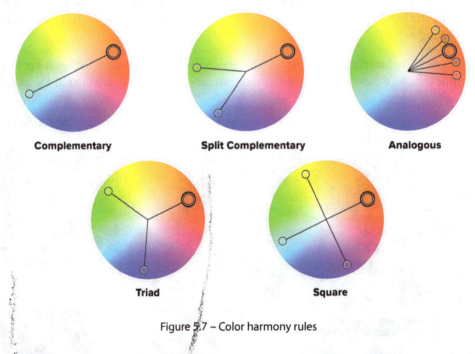

Figure 5.7 – Color harmony rules

As an activity and to further study how color harmony rules work in real time, you can check the website of Adobe Color at `https://color.adobe.com/create/color-wheel` as seen in *Figure 5.8*.

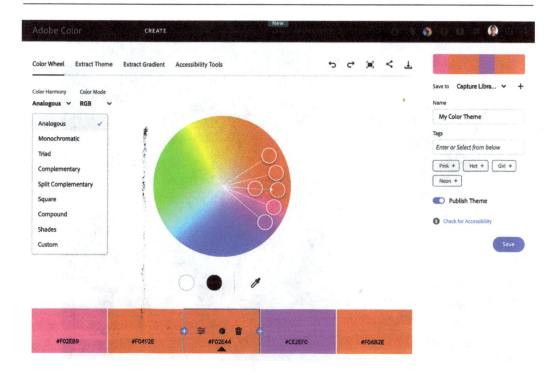

Figure 5.8 – Checking the Adobe Color website for color harmony guidance

Now that we have touched on color harmony, what is your favorite color harmony rule? Mine is **analogous**! This knowledge helps us customize our work more by adding specific colors together with the prompt, which we will see in action in the next section.

Driving color variations using color swatches

You can easily add colors that you saved in your color swatches or even create them as you go by using the color picker. Here are the steps to accomplish this process:

1. Type in your prompt. In my case, I used the previous one, which was `Salmon sushi`. This will autogenerate the color variations.

2. Head to the **Colors** section of the **Generative Recolor** panel, click the + button, and add the colors you want. This will drive the prompt to focus on those colors. You can only add five of them. See the result of adding the colors in *Figure 5.9*.

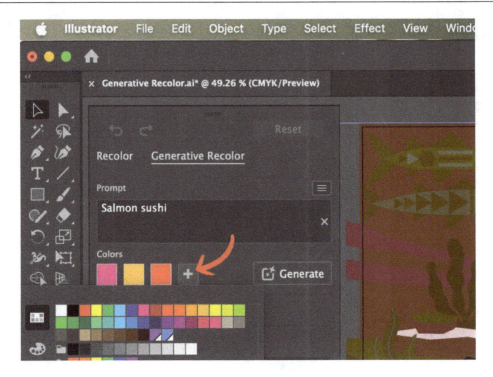

Figure 5.9 – Adding colors to drive the prompt

Now that you know these advanced controls, you also need to check how you can access the variation options. We will cover this in the next section.

Accessing more color variation options

When you generate color variations, you can hover your cursor over one of them and will immediately see a **... (see more options)** button. Clicking this gives you access to more controls, including the following:

- **Select and Edit Colors** – This enables you to gather all the colors generated and save them in your color swatches panel as a color group

- **Delete** – With this, you discard unnecessary items that were generated

- **Good Result** and **Bad Result** – These let you rate the generated result as either good or bad

You can see this in detail in *Figure 5.10*.

Figure 5.10 – Items available when clicking see more options in the variations section

Summary

Congratulations on finishing this chapter. We have learned about all of the options and methods for using Generative Recolor, accessing sample prompts, creating combinations that follow color harmony rules based on color theory, defining colors to drive our prompt, deleting variations, providing feedback in color generations, and using select and edit colors.

Generative Recolor can save us a lot of time and give us the power to create endless possibilities in our creative assets. We have now covered almost all of the modules within Adobe Firefly.

The next chapter will expand the usage of these features in your favorite desktop applications, such as Photoshop, Illustrator, and more.

Part 2:
Extending your Creative Workflow

Now that we have learned the fundamental skills of using Adobe Firefly, we will transition into using its features and implement them into your favorite productivity desktop applications and services, such as Adobe Photoshop, Adobe Illustrator, Adobe Express, and Stock.

We have also included an additional chapter that will cover other skills related to captioning text, upcoming features, sample prompts, and getting support from communities and other channels.

This part covers the following chapters:

- *Chapter 6, Accessing Adobe Firefly in Photoshop and Illustrator*
- *Chapter 7, Accessing Adobe Firefly in Adobe Express and Adobe Stock*
- *Chapter 8, Beyond Firefly*

Accessing Adobe Firefly in Photoshop and Illustrator

After learning all the necessary modules in the web browser such as Text to image, and Generative Fill, it all comes back to integrating it into our creative workflow in desktop applications such as Adobe Photoshop and Adobe Illustrator. Now, you can access these time-saving features that will boost your productivity, enabling you to get more work done in less time. Talk about combining working hard and working smart!

In this chapter, we will learn where can we access Adobe Firefly within the set of menus and commands, which will let you get comfortable with incorporating what you have learned in the previous chapters.

Specifically, we'll be covering the following topics:

- Understanding how Adobe Firefly is being implemented into Adobe Photoshop and Adobe Illustrator to create and enhance current workflows
- Incorporating Generative Fill and Expand in Adobe Photoshop (desktop and web versions)
- Using Text to Vector Graphic to create design elements seamlessly

Technical requirements

When using the desktop applications of Adobe Photoshop and Adobe Illustrator, you have to take into account the necessary processing power that you need to address to get a smooth operation in your workflow. Here are the following links for you to check if your system is up to optimal performance. Take note that you may need an active paid subscription to access the service.

Here are the Adobe Photoshop system requirements from the Adobe website: `https://helpx.adobe.com/photoshop/system-requirements.html`.

Here they are for Adobe Illustrator: `https://helpx.adobe.com/illustrator/system-requirements.html`.

If your system is within the processing requirements, then you are also good to go with using Photoshop and Illustrator on the web.

Using Adobe Firefly within the desktop application

The integration of Adobe Firefly features is intended to provide a seamless workflow both as an added feature and enhancement to existing tedious tasks you encounter, such as replacing elements, blending them together, extending or changing image orientation, reimagining color variations, and generating vector elements.

While the web browser version offers a lot of functionalities, Generative Expand is only available in Adobe Photoshop (web and desktop) and Text to Vector Graphic (Beta) and Generative Recolor are only available in Adobe Illustrator (desktop).

The Text to Template module, on the other hand, is only available in Adobe Express.

In this table, we list all of the current features with the associated access method.

Firefly feature	Web browser	Desktop Application
Text Effects	Exclusively in Adobe Express	Not available
Generative Recolor	Not available	Adobe Illustrator
Text to Image	Firefly website, Adobe Express, and Adobe Stock	Adobe Photoshop
Text to Vector Graphic (Beta)	Not available	Adobe Illustrator
Generative Fill	Firefly website, Adobe Express, and Photoshop on the web	Adobe Photoshop
Generative Expand	Photoshop on the web	Adobe Photoshop
Text to Template	Adobe Express only	Not available

Table 6.1 – Access methods for each Adobe-Firefly-powered feature

Think of using the Adobe Firefly website as a way for you to get started in using generative AI, with a simplified approach to each module that you can do easily, and one that, later on, can be implemented into other workflows that you currently have in your projects, ranging from quick designs in Adobe Express to complex compositions in Adobe Photoshop and Adobe Illustrator.

This new approach can greatly benefit new and even long-time users to adapt to using generative AI. In the next section, we will focus on Adobe Firefly features available in the desktop app version of Adobe Photoshop.

Using Generative Fill in Adobe Photoshop

Do you sometimes stare at a blank Photoshop document, not realizing that so much time has passed by because you don't have any idea what you want to create? Yes, it also happens to me, and for the most part, every designer has experienced it. Then, Generative Fill came in. This feature will enable you to create your initial compositions in minutes by telling the needed prompt as you think of it.

So, what is Generative Fill? It will enable you to add or remove anything in your canvas in a matter of time non-destructively, removing the complexities of blending and all that technical stuff needed to create that convincing composition.

But here is a disclaimer – as every tech has its hits and misses, you will encounter some of them in the process; just try and see how it works.

Let's carry out an activity by creating a composition from scratch using Generative Fill. Here are the steps:

1. Create a blank Photoshop document, which usually has a landscape orientation by default. If not, follow along using *Figure 6.1* as a reference.

Figure 6.1 – The necessary settings for creating a new Photoshop document

Using the **Rectangular marquee** tool, select the lower half of your canvas and click on the **Generative Fill** button in the Contextual Task Bar. Here, you can check *Figure 6.2* to follow along.

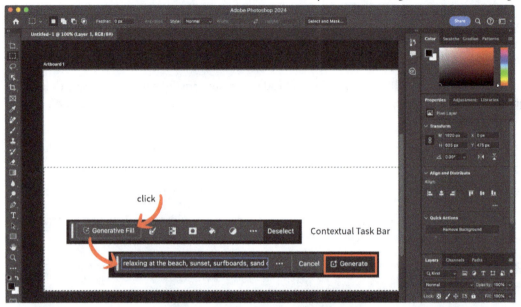

Figure 6.2 – Using the Rectangular marquee tool and typing the prompt in the Contextual Task Bar

2. Type in the following prompts and click the **Generate** button. It may take seconds or even minutes depending on the size of the selection and the prompt supplied:

```
relaxing at the beach, sunset, surfboards, sand castles with
waves
```

> **Tip**
>
> If your Contextual Task Bar is missing, you can enable it by going to the **Window** menu and making sure that the Contextual Task Bar is checked.

3. Photoshop will generate the needed image. You can see other variations on the **Properties** panel available for you to use. Notice that in the **Layers** panel, it created a new layer called **Generative Layer** (we will discuss this more in the later section).

4. Click the variation you want for it to be applied in the canvas and you can see the result in *Figure 6.3*. The result you get may vary and may not appear the same as the one seen here.

Figure. 6.3 – Using Variations and Generative Layers

> **Important tip**
>
> You can use any selection tool within Photoshop to use Generative Fill to add elements as you compose them. When removing a specific element, you can select it, leave an empty prompt, and click **Generate**.
>
> This will automatically invoke Photoshop to blend it with other layers in the process.

5.   Now that we have our initial image, we can also generate more from it without typing any prompt. Let's make a selection using the **Rectangular marquee** tool again at the remaining upper half section of the canvas.

6.   After selecting, click the Generative Fill button in the Contextual Task Bar, leave the text box blank, and click the **Generate** button. Photoshop will analyze and generate the appropriate pixels based on the open document.

7.   Select the variation that you want and refer to the sample result in *Figure 6.4* as a guide.

Figure 6.4 – Final image created without typing any prompt in Generative Fill

That is how easy it is to use Generative Fill inside of Adobe Photoshop!

> **Tip**
> This can pave the way to common industry practice for users to rapidly create images as visualized by their mind by typing them directly in Adobe Photoshop. I personally use this to skip hunting down lots of assets into stock image libraries and breaking my workflow due to jumping through browser tabs and back to Photoshop. I wish I had this much earlier before in my career.

Now that you know how to use Generative Fill, let's discuss Generative Layers, a new type of layer in Adobe Photoshop.

Understanding Generative Layers

What are Generative Layers? They are a kind of layer in Photoshop that contains the variations of the results from a requested prompt. It works non-destructively because it does not alter or modify any existing layer in your document and creates its own layer.

When you save your Photoshop document, the Generative Layer will be included in your file as embedded, thus making your file size bigger in the process. The variations generated in the Generative Layer are accessible anytime even as you close the document.

> **Tip**
> If you are short on data storage and want to optimize the file size of your Photoshop document, you can easily delete a generative variation in a Generative Layer by clicking the delete button (trash bin icon).

Here is a scenario that you can try out yourself:

1. Save your Photoshop document locally on your computer using the previous activity that we completed, preferably in your Documents folder.

2. Locate the local file in **Finder** (macOS) or **Explorer** (Windows), and check for its current size by right-clicking on the file and choosing **Get info** (macOS) or **Properties** (Windows).

3. Open the saved file in Adobe Photoshop and click the delete button on one of the variations that you didn't select.

4. Go to **File**, and click **Save**.

Check the local directory where the file resides and compare it to your previous file size. Here is an example (*Figure 6.5*):

Figure 6.5 – Size reduction when deleting variations from Generative Layers

Using Generative Expand in Adobe Photoshop

You can easily convert the image format and extend image boundaries with Generative Expand. Similar to Generative Fill, it generates pixels to suit your needs – the only difference is that you do not need to type any prompt.

Try this yourself, and follow this activity:

1. Launch Adobe Photoshop and open the previous file we create using Generative Fill, and let's extend this image. Our goal is to change it to a vertical orientation.

2. Go to the **Tools** panel and look for **Crop Tool**. You can also access it faster by pressing the C key on your keyboard.

> **Tip**
>
> Generative Expand is a feature that is only accessible using the Crop Tool.
>
> This feature is highly useful when you have images that you need to be converted into another orientation, let's say a wallpaper image into a book or magazine cover, which saves you a lot of time because it creates the necessary pixels to make it seamlessly adjusted.

3. With the Crop Tool active, drag one control point to extend the canvas using an upward motion. Refer to *Figure 6.6* to see this in action.

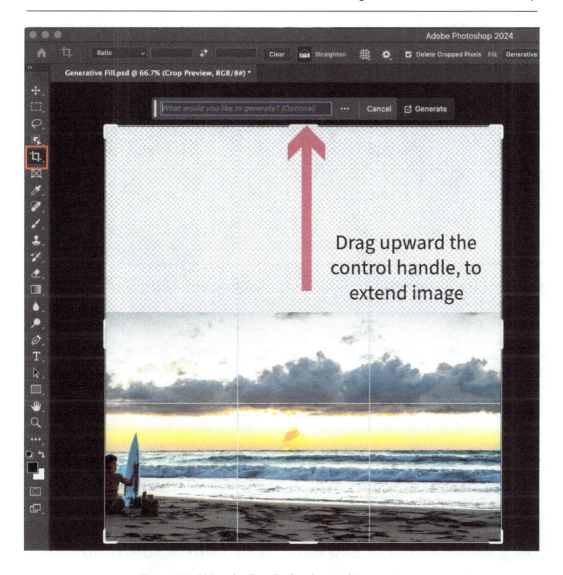

Figure 6.6 – Using the Crop Tool and extending your image

4. After extending, click the **Generative Expand** button in the Contextual Task Bar.

5. The Contextual Task Bar will change, giving you the chance to type a prompt. We will not supply anything and click on the **Generate** button. It will automatically supply you with three variations on the **Properties** panel, which you can choose to suit your needs. See the final result that I got in *Figure 6.7*.

Figure 6.7 – The final generated output using Generative Expand

That's how you can use Generative Expand in the native desktop version of Adobe Photoshop. As a bonus section, you can also access all of these features using only your web browser because Adobe Photoshop on the web also has generative AI features incorporated into it right now. Let's check how we can use it.

Using Generative Fill and Expand through Photoshop on the web

Similar to the desktop version that you need to install on your local machine, Photoshop on the Web offers you *almost* everything with only your web browser. Here are the instructions for accessing Photoshop on the Web:

1. Open your web browser (Chrome browser recommended). Type `https://photoshop.adobe.com` in the address bar and press *Enter* (Windows) or *return* (macOS) key.

2. Once the website is loaded, click the blue + icon in the top-left corner, and then click **Custom size** with all the settings needed as seen in *Figure 6.8*.

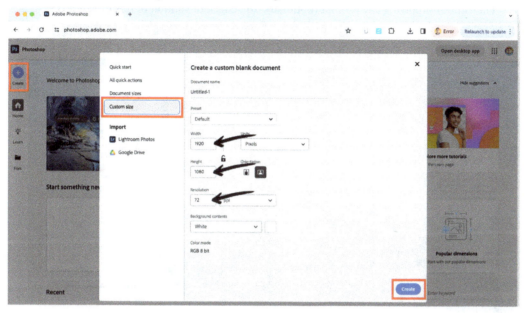

Figure 6.8 – Creating a new document in Photoshop on the Web

3. The process is almost identical to our previous activity with some changes on how you can access the **Rectangular marquee** tool. You will notice that on the web version of Photoshop, the tools are grouped into categories.

4. Using the **Rectangular marquee** selection tool located under **Select**, create a selection in the lower half of the canvas, and type the following prompt in the Contextual Task Bar:

```
underwater corals with sea horse and lion fish
```

5. It will generate three variations and a Generative Layer on which you can choose the one that suits your needs. You can apply the method of selecting the upper portion and use Generative Fill without supplying any prompt (optional). You check the output of this in *Figure 6.9*:

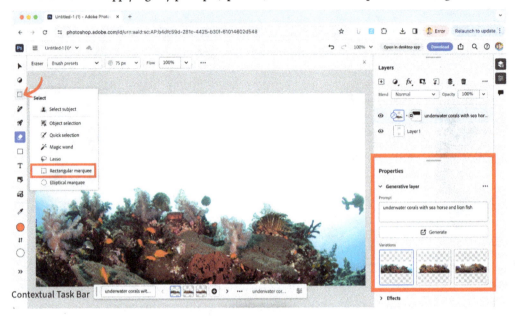

Figure 6.9 – Using the Rectangular marquee tool and Generative Fill in the lower half of the canvas

6. You can also use Generative Expand by using the crop tool located in the **Size and Position** category (the first tool in the toolbar).

    Drag upwards to extend the crop area, click the **Generative Expand** button in the Contextual Task Bar, and then click the **Generate** button and wait for the generated variations to be generated. Click the one you would like to apply. See *Figure 6.10* for the final output.

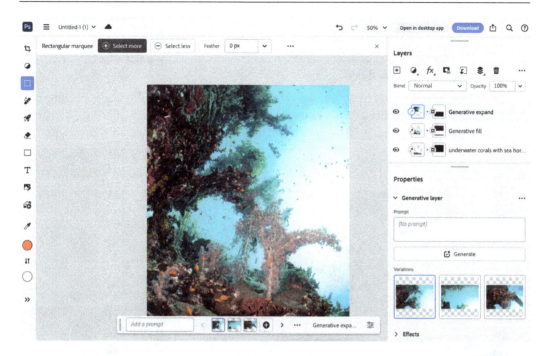

Figure 6.10 – The final output of using Generative Expand for Adobe Photoshop on the Web

That's it! You can see how the workflow was designed to easily transition from the desktop to the web version. Generative Fill and Expand are must-have skills for designers using Adobe Photoshop to ideate and make complex compositions faster than ever.

With this in mind, Adobe Illustrator also has a generative AI feature that uses Adobe Firefly, which is just as exciting to learn about and master, and that is what we will cover in the next section: the Text to Vector Graphic feature.

Using Text to Vector Graphic in Adobe Illustrator

Vector drawing can be one of those challenges that can be hard for graphic designers to master, which is why this recently released feature called Text to Vector will enable you to type in a specific prompt and it will generate the needed output for you to use. It is fully editable, scalable at any size, and has its own layer inside Adobe Illustrator.

It is a unique feature because it is not available on the Adobe Firefly web page and can only be accessed in a separate new interface in the Adobe Illustrator desktop client called the **Text to Vector Graphic** panel.

You can easily decide on the specific outcomes, such as **Subject**, **Scene**, **Icon**, and **Patterns** while blending in seamlessly in your current project for it can also have the capability to blend with the style of your illustration. We will cover these controls in the upcoming section.

Let us try and find it inside the app. We will set it up by doing the following steps:

1. Open the desktop version of Adobe Illustrator and make sure that you have the latest version of it via the Creative Cloud desktop app.

2. Go to **File**, and then click **New**. This will open the **New Document** dialog box containing all of the options you may need. In our example, let's select any blank document preset – I will be using the **Letter** document preset – and click on the **Create** button. See *Figure 6.11* for reference.

Figure 6.11 – The New Document dialog box with the Create button

3. Adobe Illustrator will open a new blank document in which we will explore where we can access the Text to Vector Graphic feature.

    By default, the classic workspace gives you access to the **Text to Vector Graphic** panel, which is located on the right side of the interface within the **Properties** panel.

Tip

In the **Properties** panel within the **Text to Vector Graphic** panel, you can click **...** to make the panel float in the interface.

You can also access it by going to the **Window** menu and clicking on the **Text to Vector Graphic (Beta)** option. You can check it using the image in *Figure 6.12*.

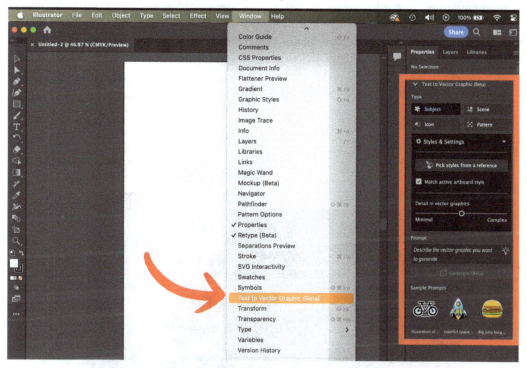

Figure 6.12 – Ways to access the Text to Vector Graphic (Beta) panel

That is how you can access the Text to Vector Graphic feature inside of Adobe Illustrator. Next, we will be creating vectors based on a prompt.

> **Tip**
>
> One important step in using Text to Vector Graphic is having a placeholder for your generated asset to reside. This is done by using any selected shape, which enables you to set a size for your output to reference.
>
> This enables you to instantly size the asset you will be generating, offering good practice in your workflow to avoid wasting time resizing elements repetitively. Think of it as using bowls and plates as you eat – everything is arranged in order.
>
> You can also click on any sample prompts and generate based on them without using any placeholder.

Creating design assets with Text to Vector Graphic

Before we create a vector graphic, we need to agree to the user guidelines stating that this technology enables us to apply styles from reference artworks, we need to have the rights to use any third-party assets, and it will also store your reference images as thumbnails along the way. This only happens the first time you use this feature. See *Figure 6.13* to get more information.

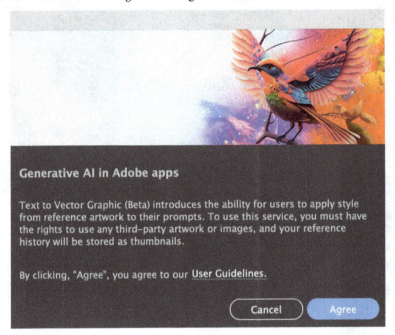

Figure 6.13 – Text to Vector Graphic notice agreement dialog box

You can easily create design assets by typing a prompt in the **Text to Vector Graphic** panel. There are some settings that you only need to take into consideration before doing so, to make your output more precise as you intend it to be. *Figure 6.14* illustrates the options.

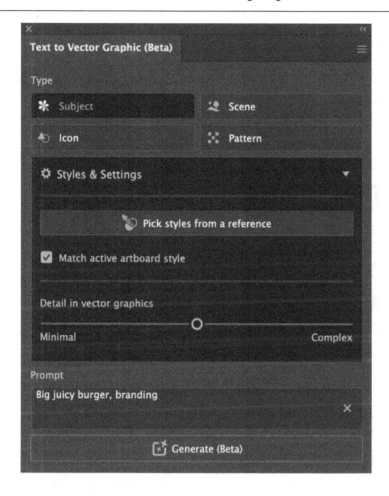

Figure 6.14 – The Text to Vector Graphic (Beta) panel

Under **Type**, we have the following controls:

- **Subject**: This enables you to create highly detailed individual illustrations, such as an object without a background

- **Scene**: This allows you to create illustrations focusing on backgrounds, typically covering the entire canvas

- **Icon**: This offers low-detail illustrations, mostly suited for logos

- **Pattern**: This is similar to **Scene**, only focusing on repeating elements

Let's try to use each one, followed by the prompt, and generate the output:

```
Subject - Big juicy burger, branding
Scene - Vintage car on a highway, desert sunset, retro poster design
Icon - Minimalist mountain
Pattern - Japanese theme waves wallpaper
```

You can see the generated output in *Figure 6.15*, but do not worry if this does not look identical to your output.

Figure. 6.15 – Generated vector graphics based on a type and prompt

After generating, let's dive into some of the advanced options, such as styles and settings.

Building advanced Text to Vector Graphic composites

Now that we know the difference between the type controls, we can easily use this to make more advanced vector designs. Using style references, we can add elements seamlessly.

While learning to create individual pieces of vector graphics is useful, in the following subsections, we will learn how to create composites. Let's try to create a vector graphic scene:

1. Create a new Illustrator file with the **Letter** preset, and create a rectangle covering the whole artboard. Make sure that this is selected because it will act as a placeholder for the generated graphic.

2. Open the **Text to Vector Graphic** panel, click the drop-down list, and select **Scene**.

3. Enter the following in the prompt text box:

    ```
    Vintage car on a highway, desert sunset, retro poster design
    ```

    Click on the **Generate** button and wait for the generated variations.

4. Select the variation you want and it will be placed in your artboard; do not worry if your resulting images don't match precisely to the one shown in *Figure 6.16*.

Figure. 6.16 – Creating a vector graphic using the Scene type

Creating a Subject type vector graphic

Creating a subject enables you to add more elements as separate independent vector drawings and is most commonly used when making compositions because it typically blends seamlessly with the background image. Follow these steps:

1. Next, let's go back to the **Text to Vector Graphic** panel and click on **Subject** in the drop-down list.

2. In the **Styles and Settings** portion, make sure to activate the use **Match active artboard style** checkbox; this will ensure that everything that we create will blend with the current style we are using.

3. Type the following in the prompt section:

```
Motorcycle rider
```

Click on the **Generate** button and wait for the generated variations.

4. Click the variation you want, and place it in your artboard; you will notice that the generated vector graphic resembles the style of the background graphic, making it much easier for designers to incorporate additional elements as needed. You can check *Figure 6.17* for reference.

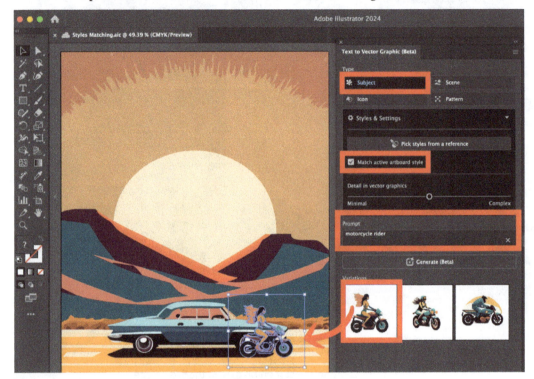

Figure. 6.17 – Adding a Subject type vector graphic with Match active artboard style enabled

Creating an Icon type vector graphic

Icons stand out when we add more eye-catching and easy-to-understand visuals. Let's try to add an icon to our project:

1. Let's select **Icon** from the drop-down list in the **Type** section and click the **Match active artboard style** checkbox.

2. Type the following in the prompt box:

        Safe travels

3. Click on the **Generate** button and wait for the generated variations.

4.   Place the generated icon in the project, as seen in *Figure 6.18*.

Figure 6.18 – Using the Icon type in Text to Vector Graphic (Beta)

Adding a vector graphic pattern

Vector graphics is something new that I discovered and it has blown my mind because it significantly helps designers ideate and implement their ideas faster than ever. Let's take a look:

1.   We can also insert a pattern in one of our designs by using the direct selection tool; let's select the background shape of the scene. Choose **Pattern** from the drop-down list in the **Type** section.

2.   Type the following in the prompt box:

```
Circular strokes
```

Click on the **Generate** button and wait for the generated variations.

3. Select the variation of choice for it to be placed in the active selection, you can definitely see the difference because of the pattern as illustrated in *Figure 6.19*.

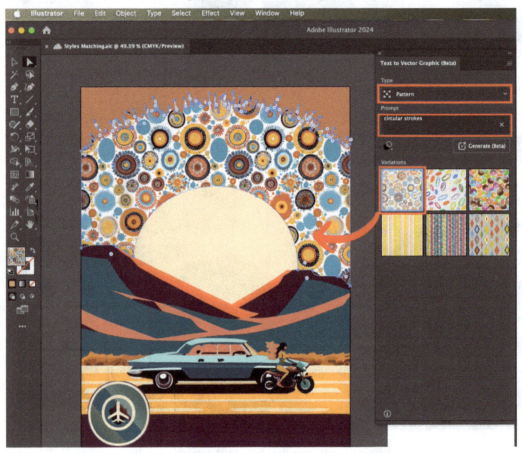

Figure 6.19 – Applying a Pattern using Text to Vector Graphic

You can now see how we created such a visually stunning design by learning how the types work in the Text to Vector Graphic feature. I hope you have felt excited and fulfilled as you have generated it on your own.

Accessing color controls in the Pattern function

The **Pattern** option is unique because it provides color options that you can use. It has three items, which are the following:

-   **Color Presets:** This enables specific adjustments in the generated pattern such as **Black and white**, **Warm tone,** and **Cool tone**

- **Limits to**: This enables you to define only a number of colors to be present upon generating the patterns

- **Specify Colors**: This lets you assign colors that you wish to be only included upon generating the pattern

See *Figure 6.20* to explore these controls.

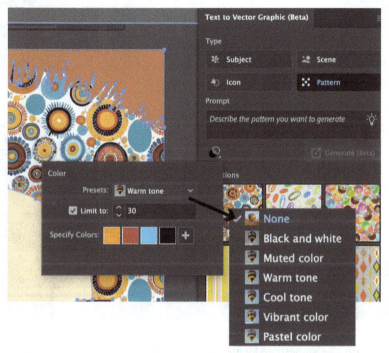

Figure 6.20 – Color control inside the Pattern option in Text to Vector Graphic

You can appreciate that this feature is powerful enough to generate vector drawing elements with the extensive options and settings you can use.

Putting it all together

Let's complete an activity in which we apply all that we learned. You will need the do the following to generate the required items in your artboard:

1. Create an empty Illustrator document with a landscape orientation; preferably **A4** or **Letter** size will do.

2. Create a scene using Text to Vector Graphic and it should reflect elements that produce an illustration of relaxing at the seaside – you have complete creative freedom to insert other elements within the canvas.

3.   You need to use at least one of each of the **Text to Vector Graphic** options, using the **Subject**, **Icon**, and **Pattern** type in the final design.

Here is the suggested visual, as seen in *Figure 6.21*, for you to get inspiration. You can also focus on other themes that you are passionate about creating as long as you create the needed element via the Text to Vector Graphic function.

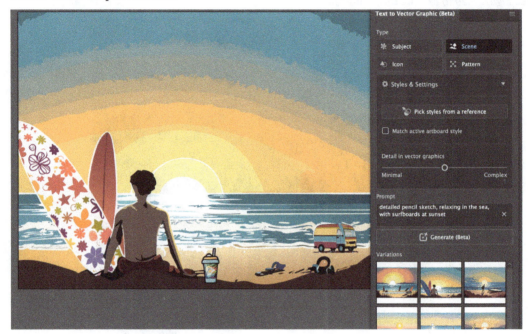

Figure 6.21 – Suggestions for implementing Text to Vector Graphic activity

Have fun and enjoy the process!

Summary

I hope you enjoyed this chapter as you have tried out your own activities using Generative Fill and Expand in Adobe Photoshop on both the desktop and web versions.

We covered how Generative Layer works and an in-depth discussion of the amazing Text to Vector Graphic (Beta) in Adobe Illustrator, as well as using the appropriate types and styles.

In the next chapter, we will focus on using more Adobe-Firefly-enabled features in Adobe Express and Adobe Stock, as we also recall the fundamental skills we have developed in the previous chapters.

7

Accessing Adobe Firefly in Adobe Express and Adobe Stock

In this chapter, we will cover how Adobe Firefly works within Adobe Express and Adobe Stock. We will discover how we can use the **Text to image**, **Generative fill**, and **Text to template (Beta)** modules within Adobe Express. Turning to Adobe Stock, we will discover how we can use the **Text to image** module to generate images that we can easily customize by using prompts that we create on our own.

In this chapter, we will cover the following topics:

- Creating visual designs within Adobe Express using the Text to image and Generative fill features
- Designing faster using Text to template (Beta) within Adobe Express
- Crafting better images with greater control in Adobe Stock using Text to image

Understanding Adobe Express and Firefly integration

Adobe Express is seamlessly integrated into each module of Adobe Firefly and takes advantage of Generative AI to create design elements faster and with ease, simply typing your prompt to produce output. As an added bonus, the images generated in Express using Firefly are safe for commercial use.

As you make your way through each chapter, you may have noticed that everything we have learned is a progression of steps that transition to integrating all of the Firefly modules together. Our goal here is to learn where we can access it all inside Adobe Express.

You can easily use the Text to image, Generative fill, and Text to template (Beta) features within Adobe Express via its intuitive user interface. Let's check out how we can do this by going to Adobe Express itself. Here are the steps:

1.   Open your browser and type in `https://express.adobe.com`.

    This will open the Adobe Express website, into which you will need to log in using your Adobe ID. After logging in, you will see the interface shown in the following figure.

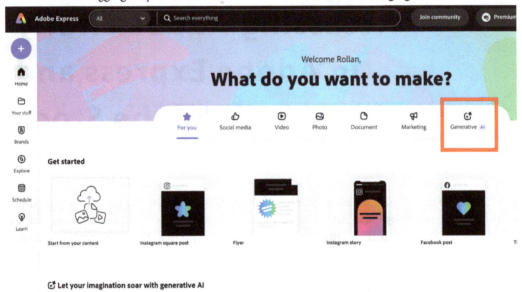

Figure 7.1 – The Adobe Express web interface with the Generative AI feature highlighted

2.   To access all of the Adobe Firefly-related features, you can click on the **Generative AI** category and it will display a new set of items as seen in *Figure 7.2*.

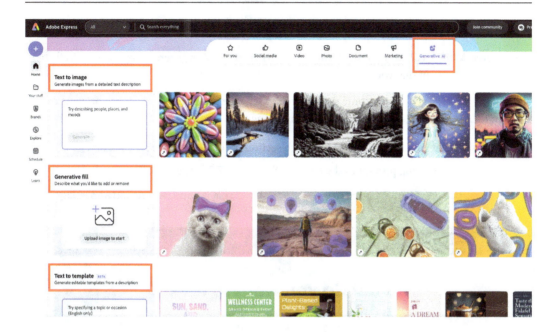

Figure 7.2 – Adobe Firefly-enabled modules within the Generative AI category of Adobe Express

3. Scroll down when needed to see all of the available modules within Adobe Express. It also offers you the chance to type in your prompts on this screen to get started quickly.

> **Tip**
>
> To get started even faster, you can hover your cursor on the image samples on the right to get some insights on what specific prompt was used, (in the Text to image, Text to template (Beta), and Generative fill modules), you will see the before and after results.

Now that we know where we can access the main page of all the related Generative AI features of Adobe Express let's learn to use Text to image within Adobe Express.

> **Important tip**
>
> Firefly modules are readily available within Adobe Express, and you don't need to navigate to the Generative AI category on the website to use them. This feature allows you to unleash your creativity anytime while using Adobe Express.

Using Text to image in Adobe Express

As we covered in the basics in *Chapter 2*, you can easily create images generated by Adobe Firefly by typing a specific prompt that you need.

As an activity, let's try generating some decorative digital wallpapers that you can use on your own computer. Sounds like fun, right? Come and follow along with me:

1. Make sure that you are in the **Text to image** section of the **Generative AI** category.

2. Type in the following prompt in the textbox. See *Figure 7.3* for guidance:

   ```
   3D splash with irregular forms made from translucent glass,
   textured gold, cotton, red velvet silk
   ```

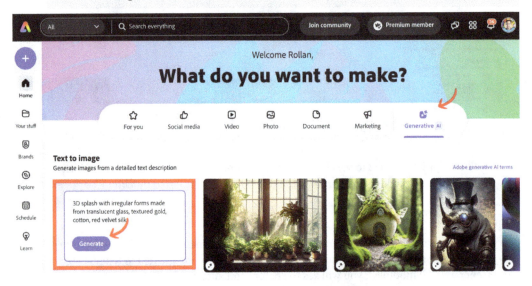

Figure 7.3 – Accessing the Text to image module within the Generative AI category

3. Click the **Generate** button to process your request and wait for the output to appear. Do not worry if the size of the canvas is square for now; we will change it later.

4. Click on the **Content Type** dropdown and change it to **Art**. Click **Generate** again to recreate the output based on the changes we made. Do not worry if the output shown in *Figure 7.4* does not directly resemble yours.

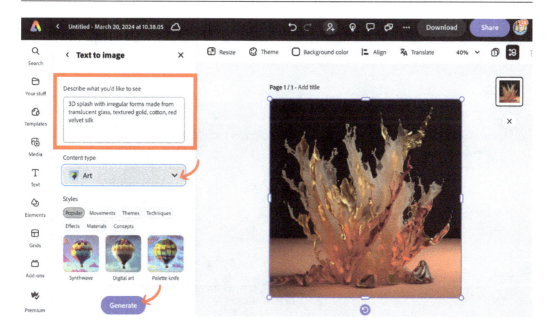

Figure 7.4 – Changing the Content type and generating new images

5.   As an activity, feel free to change the **Styles** setting based on your preferences. You still have the option to choose one variation from resulting list of the output images . Next, we will resize the canvas to apply a **16:9** aspect ratio, which is perfect for digital wallpapers.

6.   Click **Resize** on the top portion of the Adobe Express interface, which will open the **Resize** options In the left panel. scroll down and find the **Photo** category.

7.   Click the checkboxes for **Desktop wallpaper** and **Phone wallpaper** and then click the **Duplicate & resize** button located on the bottom left of the interface. See *Figure 7.5* for reference.

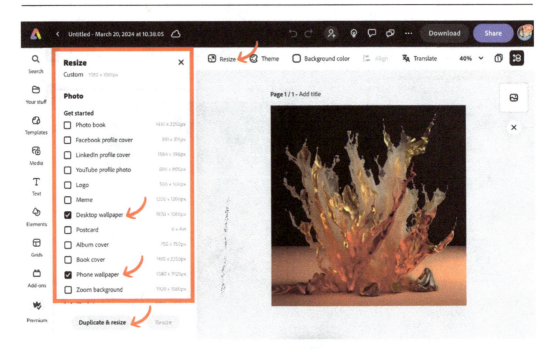

Figure 7.5 – Using the Duplicate & resize feature in Adobe Express

8.   Adobe Express will create copies of the generated image that you can adjust on your own. As seen in *Figure 7.6*, we now have three options showing: our original square image, a desktop wallpaper, and a phone wallpaper.

Figure 7.6 – The result of creating multiple images via Duplicate & resize

9. As an activity, double-click the second option, which is our desktop wallpaper. Click the image (if not already selected) and adjust the control handles of the Text to image output to fill the canvas size. See *Figure 7.7* for reference.

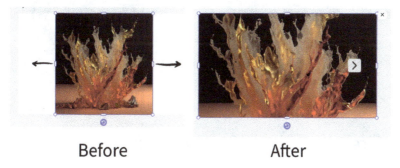

Figure 7.7 – Dragging the control handles of the Text to image output to fill the canvas

10. When you are done with making adjustments to the image, click the **View all pages** button on the top-right section of the interface to switch to page three and adjust that one.

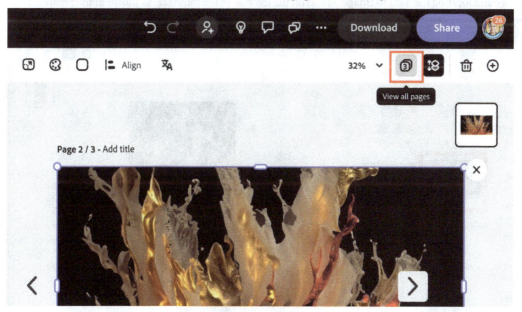

Figure 7.8 – Using the View all pages button to switch pages

11. I will let you figure it out on your own regarding the phone wallpaper canvas. Broadly, we use the same procedure, only dragging the vertical controls rather than horizontal handles as done in *step 9*. It should look similar to the following figure.

Figure 7.9 – Recommended adjustments for the phone wallpaper image

If you change your mind regarding your prompt at any stage, you can change things up by clicking the image and locating the **Text to image** button on the left side of the interface, as shown in the *Figure 7.10*.

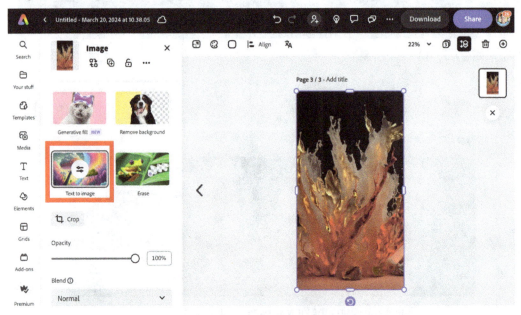

Figure 7. 10 – Editing your prompt using the Text to image button

Important tip

You can easily access and add more Text to image functions by clicking **Media**, and then the **Text to image** button. This enables you to type in a prompt and hit **Generate** again, allowing you to stack more layers in the process of creating the best project possible. See *Figure 7.11* to see it in action.

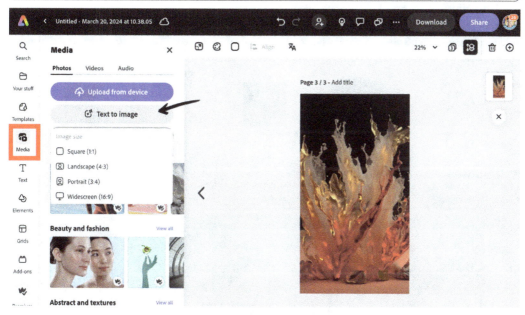

Figure 7.11 – Adding more Text to image layers to your project

There you have it! You now know how to use Text to image inside Adobe Express. With this feature, you can easily create amazing backgrounds and photorealistic images to use in any project you have to tackle.

I personally use it for creating unique-looking design features such as presentation backgrounds, promotional posters, social media posts, Zoom backdrops with client branding, and overlay graphics for live streaming – the list goes on.

Now that we have covered creating images, let's dive in a little deeper into altering our work using Generative fill in the next section.

Editing faster with Generative fill in Adobe Express

In *Chapter 3*, we discussed in depth how Generative fill works on the Adobe Firefly website. Using it in Adobe Express is such a breeze that it will be of second nature to you when you want to use it.

You can access Adobe Firefly features in Express by going to the homepage and scrolling down till you see the **Generative fill** section as shown in *Figure 7.12*.

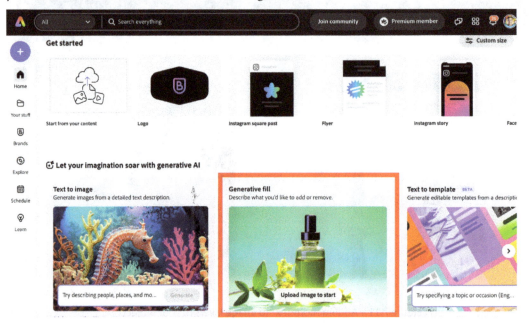

Figure 7.12 – Accessing Generative fill within the Adobe Express home page

Let's edit the sample image provided by adding and removing elements. You can follow along with these steps:

1. Click on the **Generative fill** option to enable the Generative fill workspace. It will open an image in which we will remove everything except the green bottle.

2. Using the brush, you need to make a big selection covering almost all of the areas to be removed. Remove any text from the prompt text box to make sure that we will not generate anything in the process. Adobe Express will analyze the image and remove the selected components seamlessly. Check how I did the selection in *Figure 7.13*

Figure 7.13 – Brushing over distracting elements in the photo to remove them

3. Click the **Generate** button and wait for Generative fill to show the results that you can choose from.

4. Click the first result and then click the **Done** button. I tried this procedure multiple times with other image assets and it always gives me the intended output, as seen in *Figure 7.14*.

Figure 7.14 – Result of removing items using Generative fill

Generative fill inside Adobe Express offers a much simpler user interface than the equivalent on the Firefly website, which has lots of settings. One practical use of Generative fill is to do quick editing of projects inside Adobe Express by only using the web browser.

That is how easy it is to use Generative fill inside of Adobe Express! You can access this feature whenever you have an image actively selected and look for the **Generative fill** button on the top left of the interface, as shown in the following figure.

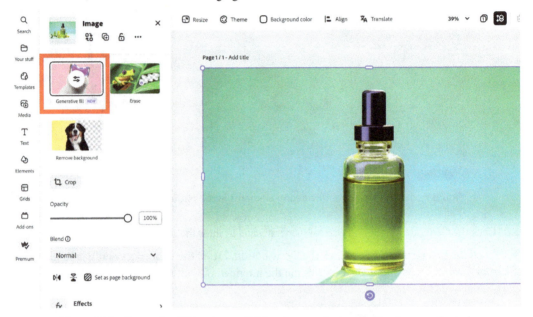

Figure 7.15 – Generative fill button available when you have an active image selected

As an activity, let's add more items to the image using Generative fill. Feel free to add any items that you would like to incorporate – my goal is for you is to exercise your creative muscles. The following figure shows the final image that I made.

Figure 7.16 – Suggested final image of the activity with Generative fill in Adobe Express

Can you just imagine how much time can you save using this feature, especially when you need to deliver assets on a rushed deadline? With that in mind, let's continue to level up our content creation process by learning the next Adobe Firefly feature, Text to template (Beta).

Generating designs using the Text to template (Beta) feature

There will be times when creative burnout will hit you, especially when you are designing all day with Adobe Express or any other design tool available on the market right now. When this happens, you can also take advantage of real-time collaboration in Adobe Express so you can work together with your colleagues. With that in mind, even if you are the only one doing the design work, it is good to know that you have the **Text to template (Beta)** feature to help.

This feature is only exclusively available in Adobe Express, even if you are using the free plan. So, what is Text to template (Beta)?

Text to template (Beta) enables you to generate fully editable design templates in seconds by simply typing the prompt that you want. Sounds promising, right? Let's learn how to use it with the following steps:

1. Open your web browser and go to the Adobe Express website at `https://new.express.adobe.com/`. Then, scroll down and click on the **Text to template (Beta)** section, as seen in the following figure.

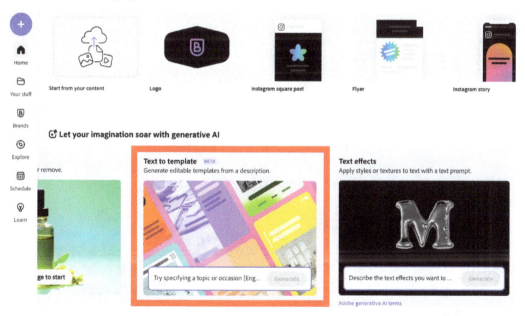

Figure 7.17 – Accessing Text to template (Beta) from the Adobe Express website

2. This will open a floating pop-up dialog box with a few options, including template sizes, prompt suggestions, and ready-made AI-generated templates. See the following figure for some examples.

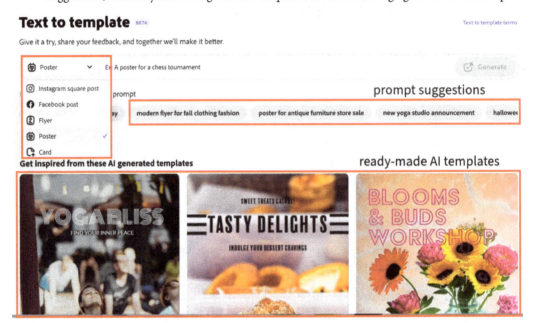

Figure 7.18 – Text to template (Beta) controls

3. As an activity, let's create a birthday card by choosing the **Card** option in the size dropdown. Type the following prompt in the text box:

```
Colorful pink themed, 3rd birthday party with unicorn cake with
sparkling candles
```

Click the **Generate** button and wait for the processing to finish. See the following figure for our results:

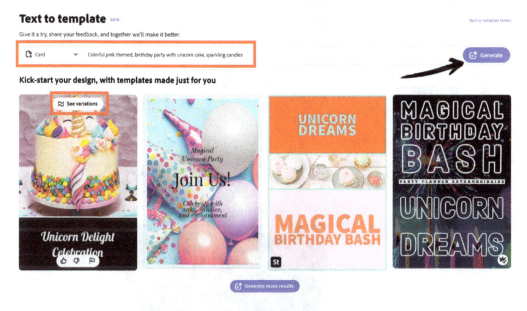

Figure 7.19 – Generate templates using Text to template (Beta)

4. Click the template that takes your liking. Doing this will open another tab in your browser with the template fully editable in Adobe Express, as shown in the following figure.

Important tip

To take full advantage of the Text to template feature, it is advised to have an existing Premium plan that will cover the requirements for safe commercial use.

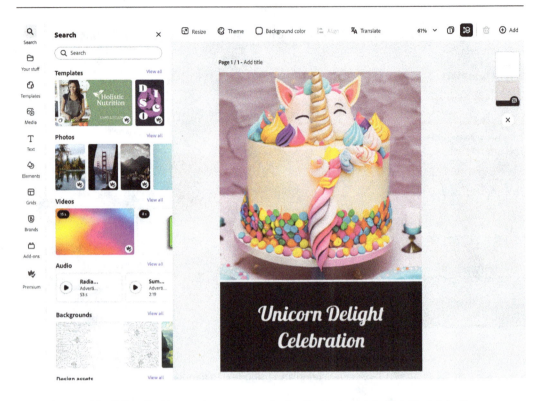

Figure 7.20 – Fully editable template generated using Text to template (Beta) in Adobe Express

> **Tip**
>
> When you generate a template that is almost exactly what you wanted, you can hover over the template and click **See Variations**. This will enable you to fine-tune the method of generating templates with a close resemblance to your initial pick.

There you have it! Generating designs using Adobe Firefly with Text to template is simply magical and saves you lots of time that you can now spend on focusing and customizing the template to fit your needs.

Speaking of customization, you may want to check the next section on how to find the perfect photo to add to your designs. You can find this with Adobe Stock.

Generating images using Firefly in Adobe Stock

I don't know about you, but most of the time I spend doing creative work is on finding that perfect image that will bring your idea to life. What if you could generate that sought-after image from within Adobe Stock itself, which also has the Adobe Firefly Text to image module? You can do this easily with the following steps:

1. Open your web browser and navigate to `https://stock.adobe.com/generate/`.

2. The website enables you to generate images from Adobe Stock using the Text to image module of Adobe Firefly, as shown in the following figure.

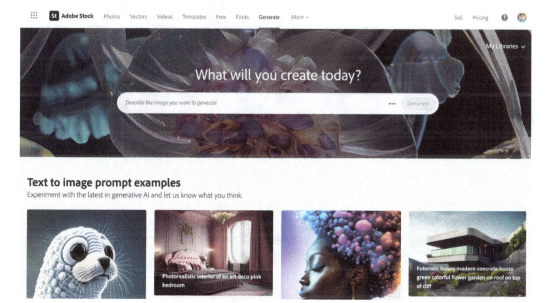

Figure 7.21 – Adobe Stock website where you can generate images

3. This is a much more simplified version of Text to image with fewer options to choose from. As an activity, let's try to generate something using the following prompt:

    ```
    onigiri, harumaki, ramen, donburi, separate dishes in table with
    japanese restaurant ambiance photo, studio light, vibrant colors
    ```

4. I added additional parameters such as changing the aspect ratio to **Widescreen (16:9)**, setting **Content type** to **Photo**, and choosing **Hyper realistic** for **Style**. Once you've made your choices, click **Generate** at the bottom of the interface. Click each thumbnail to see it in a larger view and download it as a JPEG file. You can see the generated images that I got in the following figure.

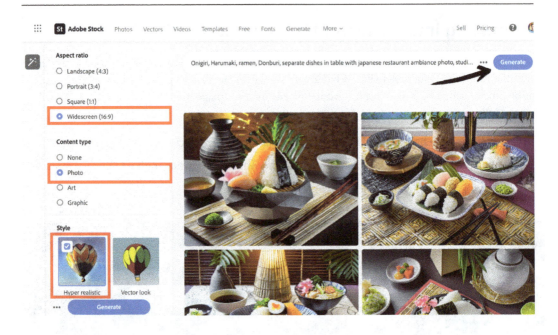

Figure 7.22– Using Text to image inside Adobe Stock to generate images

> **Important note**
>
> You can take advantage an Adobe Stock subscription to download the images created with the Text to image feature. You can check out a 30-day trial together with other suitable plans that fit your workflow at `https://stock.adobe.com/plans`.

I prefer to use the Text to image feature on the Adobe Firefly website as it offers the latest Model implementation that generates more accurate results, along with offering better ease of use and the ability to download the generated images without any paywall. Having your images organized in Adobe Stock does offer a one-stop shop for managing all of the creative assets you need in an orderly manner, though.

Summary

In this chapter, we have covered a lot of Adobe Firefly features integrated into Adobe Express and Adobe Stock. We created images using Text to image, provided ways to edit images faster using Generative fill, and had lots of fun generating fully editable templates using Text to template (Beta). We also saw how to use the Text to image feature directly in Adobe Stock.

I want to thank you for reading this chapter, and I hope you gathered lots of insights and information that you can use to confidently navigate, create, and publish works using Adobe Firefly. I'm looking forward to sharing more in the final chapter of this book, as we cover more interesting topics on Firefly and beyond!

8

Beyond Firefly

We are now in the last chapter of the book, and I want to thank you for being on this learning journey with me in using Adobe Firefly, together with other Adobe Creative Cloud desktop applications; this certainly has pushed the limits of digital creativity further and will continuously do so, beyond the use of generative AI.

We will cover more on the recently announced technologies that are within Adobe Firefly and the upcoming features that have been publicly announced in beta form, such as Text to Texture (Sampler) and Generative Background (Stager Beta) in Adobe Substance 3D.

We will provide you with some techniques that will help you, such as reverse prompt engineering and a cheat sheet that you can quickly refer to when you need to refresh your knowledge.

In this chapter, we will cover the following topics:

- Discovering the latest Firefly-related affiliations and partnerships
- Learning about the Adobe Firefly-enabled features in Adobe Substance 3D
- Accessing techniques, such as reverse prompt engineering and the cheat sheet

Firefly-related affiliations and partnership

My aim in this final chapter is to get you started in learning more than what this book has to offer, challenging yourself to be on the cutting edge of this fast-paced world of technological changes.

I can still remember the days of Adobe Creative Suite in which you had to wait for a 12–24-month cycle to get a new version of the software in a CD format. Technology has really changed; now, you can experience an update of the latest version in just a month!

Way back in *Chapter 1*, as we started to learn Adobe Firefly, we mentioned a few affiliations with other generative AI players in that industry that helped shape what we now experiencing in using Adobe Firefly. Let's take a deeper look at how significant this topic is, starting with Google Bard (now Gemini).

Google Bard (now Gemini) was unveiled on May 10, 2023 at the Google I/O conference event, in which Google announced an association with Adobe Inc. to allow for more ethical artificial image generation.

Google and Adobe are teaming up to make Firefly available in the Gemini chat window. Google commits to respecting and protecting content creators, so images made with Gemini will use Firefly's Content Credentials technology.

This tech lets creators link information to their images, helping them prove their work's authenticity or any changes they've made to it. Unfortunately, in February 2024, after Bard was renamed Gemini, it stopped using Adobe Firefly as its default text-to-image generator. Google's steering committee members together with Adobe have formed another movement called **Coalition for Content Provenance and Authenticity (C2PA)**. You can learn in-depth information about C2PA using this link: `https://c2pa.org/faq/`.

Similar to the partnership between two industry tech giants Google and Adobe, NVIDIA and Adobe shared a common vision to empower creators with powerful AI tools while ensuring responsible, ethical use of technology, NVIDIA being a featured member of the Content Authenticity Initiative. The result was **NVIDIA Picasso**.

NVIDIA Picasso provides a cloud service for users to create and deploy applications powered with generative AI. Adobe has provided access to some of the models. In return, Picasso's high-performance infrastructure provides high-quality outputs for delivery in Adobe Firefly and other Adobe Creative Cloud applications.

Adobe Experience Manager (**AEM**) is closely related to Adobe Firefly in that it has been the platform on which innovation has been implemented as a priority.

AEM is a comprehensive content management solution that enables enterprises to easily deploy digital touchpoints, such as a website, mobile app, and forms, into real-world campaigns, driven by data, analytics, and so on.

Using AEM Assets, which is tightly integrated into Creative Cloud desktop applications with Adobe Express and Firefly, you can deliver better asset management and scale significantly. To learn more, here is a video explaining all of the details: `https://www.youtube.com/watch?v=XUrOG8LFPSo`.

Let's now put our thinking hats on and explore the new features that are now in beta form in Adobe Substance 3D.

What is Adobe Substance 3D?

Adobe Substance 3D is a collection of advance 3D applications and services that were acquired by Adobe Inc. way back in 2019 through Allegorithmic; you can check out more detailed information at this link: `https://www.adobe.com/creativecloud/3d-ar.html`.

Visiting this that link will open up another dimension of learning, focusing on highly advanced 3D imagery techniques, such as modelling, effects, painting, blending, texturing, and authoring augmented reality projects.

You only need to dedicate time and focus on completing the guides and tutorials and practicing your own projects, enabling your current design skills to reach the next level, from being a print/digital designer to adding a new skillset to become a multidisciplinary 3D artist!

This is where I discovered my love for Stager, which has enabled me to land projects, conduct numerous series of workshops from 2018 to 2020, and greatly benefit me to position myself as a pioneer in using Adobe 3D tools in my country.

Watch the tutorials because they are well structured, with helpful sample files that make it easy for you to understand and apply.

Now, in the next section. let's learn the necessary system requirements to run Adobe Substance 3D smoothly.

The system requirements for Adobe Substance 3D

When using Adobe Substance 3D, you have to take into consideration that producing high-quality images will require lots of computational processing power, to achieve the necessary output to be delivered in the best experience possible. Check out the following link to see whether your system is compatible with Adobe Substance 3D: `https://www.adobe.com/products/substance3d/discover/hardware.html#minimum-requirements`.

You may also need to use an **active paid subscription** or take advantage of the **30-day trial** opportunity.

Adobe Substance 3D Firefly-enabled features

Adobe Firefly is going to be added to Adobe Substance 3D, and it is offers time-saving features that will enable you to work faster than ever with the use of these two generative AI technologies. Adobe Substance 3D's Firefly- enabled features show promising results to boost your workflow. let's take a look and explore how we can do it ourselves in the next section.

What is Adobe Substance 3D Stager (Beta) and Sampler?

Often known as **Stager (Beta)**, this platform allows you to create realistic 3D visualization images without learning how to master complex 3D applications, making it suitable for non-designers. With that said, Adobe Firefly has been integrated into Stager (Beta) with a feature called **Generative Background (Beta)**.

Similarly, another Substance 3D app called **Sampler** enables you to create 3D textures, materials, and lighting setups by capturing them in the real world using a simple scanning device, such as a camera (photogrammetry). It also has an Adobe Firefly-enabled feature called **Text to Texture (Beta)**.

You can access this by installing the *beta* version of Stager and Sampler in the Creative Cloud Desktop app. Follow the following steps:

1. Open the **Creative Cloud** desktop app.

2. On the left side of the interface, look for the **Beta apps** category and click it.

3. You will see all of the available apps currently in beta; look for **Substance 3D Sampler (Beta)** and **Substance 3D Stager (Beta)**.

4. Click the **Install** button on the right side and wait for the process to completely install Sampler and Stager (Beta).

> **Note**
>
> Beta applications can often be unstable and stall the process of improvement and testing. Your experience with this may differ from an official build released in the future.

You can see *Figure 8.1* as your guide on how to install the beta versions in detail.

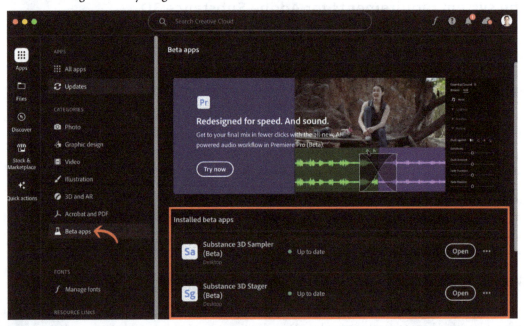

Figure 8.1 – The Adobe Creative Cloud desktop app – selecting the Beta apps category

I have installed both of them on my computer, so let's get started using those Firefly features.

Using Generative Background (Beta) in Stager (Beta)

With Generative Background (Beta), you can input a prompt that will enable you to rapidly create background images that align with the camera perspective and lightning of your objects in 3D space, using Match Image technology.

Let's take a look at how we can use it:

1. Open **Stager (Beta)**, go to the **Starter Assets** panel, and choose any of the models available (**Cream tube**, **Bottle round**, and **Jar twist small** are the ones I used.

2. Type the following prompt into the **Generative Background** panel (on the left side of the interface):

    ```
    Natural wood product display in a natural tone studio. Foliage
    in background.
    ```

3. Use the following settings for each option:

    - **Aspect Ratio: Widescreen 16:9**

    - **Color and tone: Vibrant colors**

    - **Lighting: Studio light**

    - **Composition: Closeup**

4. Click the **Generate** button and wait for the result to process.

5. Click the one that suits your taste; doing so will automatically activate the Match Image feature to fix the lighting, perspective, and camera settings.

Like most recent technologies, the result of Generative Background (Beta) may or may not suit your needs. At my end, I applied some basic transformations and scale operations to suit my needs as I explored how it works.

Here is the suggested result that was generated (*Figure 8.2*):

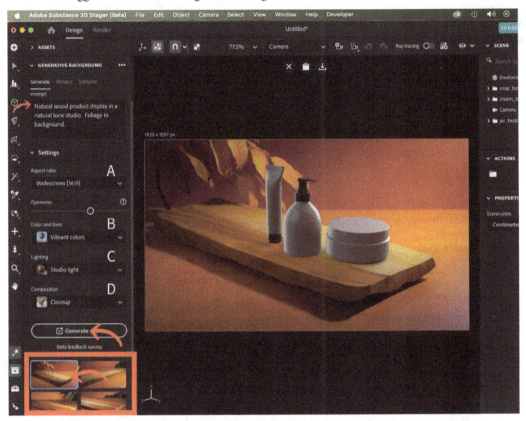

Figure 8.2 – Using Generative Background (Beta) in Adobe Substance Stager (Beta)

To export the project into an image file, click the **Render** mode at the top of the interface, set it with a **Medium** setting in the preset, and click the **Render** button. Wait for the render to finish.

I used the **Medium** preset because it is the perfect balance of getting high-quality images rendered without spending a day just to see the final output. As a comparison, using the **Medium** preset required me to wait for four minutes, while using the **High** preset to render the same project took 27 minutes. Check out *Figure 8.3* for more guidance on how to render.

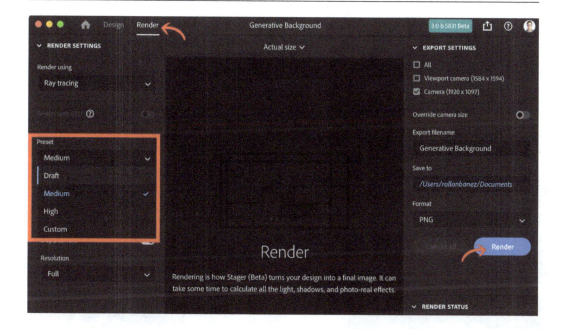

Figure 8.3 – Rendering your final image together with the settings needed

That is how you use Generative Background (Beta); it gives you the power to utilize your prompt engineering skills, saves you time finding the right image for your background, while solving concerns such as inconsistent lighting, shadows, and composition arrangements that would take even seasoned 3D professionals lots of time to execute.

It is the Text to image feature in Adobe Firefly, plus the ease of using 3D objects, that allows you to produce great images in minutes. It truly delivers fast 3D product visualization at a level of productivity unseen before the use of Generative AI.

Next, we will learn how to use Text to Texture in Sampler.

Using Text to Texture (Beta) in Sampler (Beta)

Creating textures to generate realistic-looking 3D assets is a challenging task, especially if you are starting from scratch with Adobe Photoshop as your main tool, with 16+ tabs in your browser just to find stock images and deadlines looming closer. That alone is just the selection process, and you have not even reached the stage of applying the texture itself to the actual object you want to preview to see whether it works or not.

Let's try out Text to Texture as a way for you to work faster and achieve higher-quality results in a faster time using Adobe Firefly.

Follow the following steps to try out Text to Texture:

1. Open the **Sampler (Beta)** application and click on the **GENERATIVE (BETA)** panel. You will see that there is a sphere present by default that you can apply the texture to. See *Figure 8.4* for guidance.

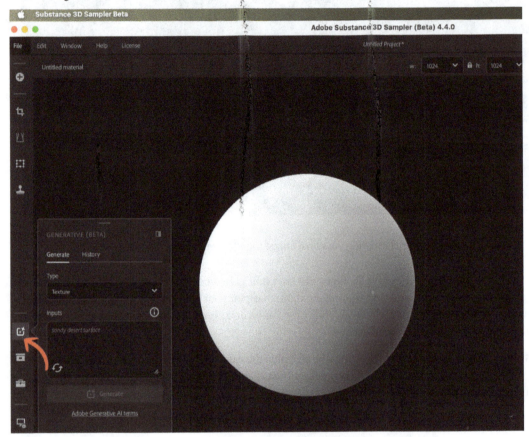

Figure 8.4 – Accessing the Text to Texture (Beta) panel in Sampler (Beta)

2. Type in any prompt you wish to generate as a texture; here are some examples to get you started:

    ```
    Cobblestone ground with pebbles
    Carved wood with burned charcoal
    Smooth glossy square ribbed plastic
    ```

3. Click the **Generate** button for Firefly to process your request.

4. In the **Results** section of the **GENERATIVE (BETA)** panel, drag and drop any of the results into the sphere at the center of the canvas (see *Figure 8.5*):

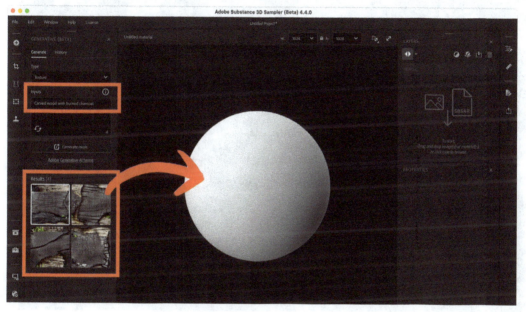

Figure 8.5 – Applying the Text to Texture result

5. The **MATERIAL CREATION TEMPLATE** dialog box will then pop up. Choose **Image to material | AI Powered**. Click the **Import** button located at the bottom right. See *Figure 8.6* for guidance.

Figure 8.6 – The MATERIAL CREATION TEMPLATE dialog box

6. Wait for the process to finish, as it will apply the generated texture to the base material. You can see the final result of this process in *Figure 8.7*.

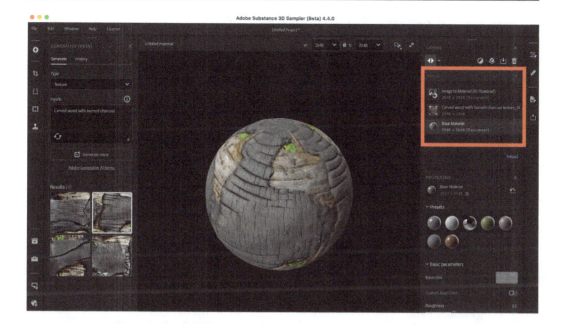

Figure 8.7 – The successfully applied Text to Texture (Beta) texture

You can customize it more if you wish. Have fun exploring Text to Texture; it will definitely be a time-saving workflow for game developers, or any 3D artist who deals with asset generation as part of their job, once it is officially released and approved for commercial use. I look forward to these features to improve more in the near future.

> **Important note**
> Beta features and applications are only for exploration and testing purposes for product improvement; they are not recommended for commercial purposes.

Let's cover some upcoming Firefly features in the next section.

Upcoming Adobe Firefly features

Since Adobe Firefly was launched in March 2023, Adobe has given us a sneak peek into some promising features it will implement in the platform. Although we should not assume all of it will be available going forward, you can never tell what the future holds.

You can check the officially released videos on the upcoming features here:

- *Adobe Firefly: Future Explorations*: `https://www.youtube.com/watch?v=_sJfNfMAQHw`

- *Future Vision: Firefly for @AdobeVideo*: `https://www.youtube.com/watch?v=30xueN12guw`

- *Generative AI in Premiere Pro powered by Adobe Firefly*: `https://www.youtube.com/watch?v=6de4akFiNYM`

A new set of generative AI features in Photoshop (Beta) has been added in latest update on April 24, 2024. A few of these features are Generate Image, Reference Image, Generate Similar, Generate Background, and Enhance Detail. This update arrived on the heels of the previous update in March 2024!

I created a table that lists some of the features that were publicly available already at the time of writing, some of which are still unavailable . These features were announced at the Adobe MAX conferences, as well as through Adobe Firefly website publications and videos officially approved by Adobe Inc. on its social media channels.

Feature name	Status
Text to Image	Available
Generative Fill	Available
Text to Template	Available
Generative Recolor	Available
Extend Image (Generative Expand)	Available
Text to Vector Graphic	Available
Text Effects	Available
3D to Image	Coming soon
Conversational Editing	Coming soon
Sketch to Image	Coming soon
Combine Photos	Coming soon
color-Conditioned Image Generation	Coming soon
Upscaling	Coming soon
Project Stardust	Coming soon
Generative Audio and Video	Coming soon
Custom Models (Enterprise only)	Coming soon

Table 8.1 – Availability of features in Adobe Firefly

> **Note**
> At the time of writing, Adobe has not provided an official list of features coming to Adobe Firefly, and this table is subject to change in the future.

Project Stardust is one upcoming project that will enable you to redefine how you do photo editing right within a browser. It enables you to work without layers, manipulate selections using object detection, and generate images using generative AI. Get to know more about it using this link: `https://labs.adobe.com/projects/stardust/`.

As mentioned in *Chapter 1*, features such as Generative Fill, Text to image, Text Effects, and Generative Recolor are all commercially available modules that were trained exclusively using Adobe Stock images, openly licensed content, and public domain content, making them safe for commercial use.

Adobe Stock Contributor Program is a program with a compensation model that allows creators to thrive in monetizing their contributions through the assets that they submit and are approved by the moderators. You can check in-depth information about this program at `https://contributor.stock.adobe.com/`.

Now that we have this table to guide us on what Firefly features we can get our hands-on currently, this chapter wouldn't live up to its name, *Beyond Firefly*, if we didn't cover some more advanced features to take prompt engineering to the next level. It is time to learn how we caption images.

Reverse engineering prompts – captioning images

You may recall that in *Chapter 2*, we covered a section on how to write effective prompts using guide questions and statements to achieve better results in the Text to image module. Now that we have acquired the skill of prompt engineering, let's look the other way around by learning a technique to reverse-engineer prompts. It is called **captioning images**.

Captioning images enables you to easily identify how to decode images into words, together with specific styles that they may contain. You can also look at this technique as a program being deconstructed into code for you to recompile and make iterations. Another benefit is having the capability of an art historian with years (or even decades) of experienced practice to decode detailed information about a digital image, right at your fingertips.

One way you can caption images is by using an online tool (model) called **Img2Prompt**, made by a company called **Replicate**. It is publicly available and easy to use as a learning tool to discover prompts/words (see *Figure 8.8*).

You can check it out using this link: `https://replicate.com/collections/image-to-text`.

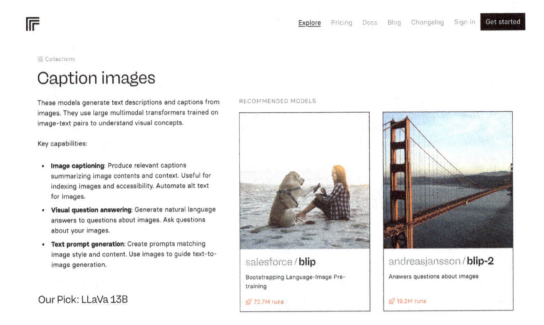

Figure 8.8 – A collection of tools available to caption images via the Replicate website

In that collection, there are lots of caption imaging tools, and we will focus on Img2Prompt because it gives appropriate text prompts, including styles associated with the image, and, in my experience, offers a lot of suggestions for learning and discovering great artists.

As an activity, our goal is to use Img2Prompt to identify elements of a photo and provide prompts that will sufficiently describe the image. You can use a publicly available photo of yours, as I did myself, and let Img2Prompt provide captions for my image using the following steps:

1. Open your browser and go to this link: `https://replicate.com/methexis-inc/img2prompt`.

2. Drag and drop your photo (on Finder/Explorer for Mac/Windows, respectively) to the image file section of the website (the lower section with the trash bin icon).

3. Click the **Run** button and see the output results in the right section; it contains the captioning done to the image. Note that it may introduce some artist names in the process because Img2Prompt was optimized for Stable Diffusion.

See *Figure 8.9* as a guide to do this experiment.

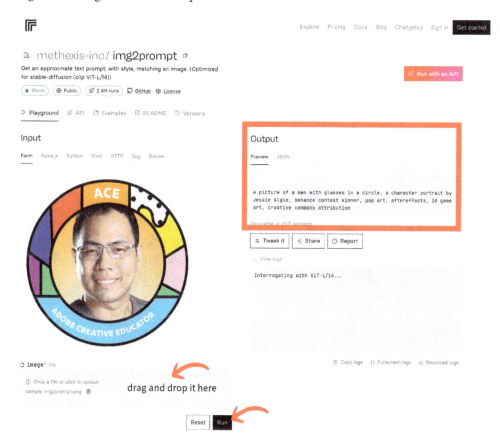

Figure 8.9 – Using the Img2Prompt tool to caption an image and detect its content in prompts

The goal of this activity is to explore how these tools (models) behave when creating captions to images, within the range of acceptable accuracy, to detect subjects in an image. One suggestion is to try the generated prompts in Firefly, removing any mention of an artist's name.

Adobe has been very strict in considering ethics and implemented a set of user guidelines for the use of Adobe Firefly and other generative AI products. You can see the details using this link: `https://www.adobe.com/legal/licenses-terms/adobe-gen-ai-user-guidelines.html`.

With that in mind, let's move on to the next section, in which we will explore valuable resources and our cheat sheets.

Exploring the prompt cheat sheet

This prompt cheat sheet is a part of my personal collection of generated images in Adobe Firefly. Feel free to mix and match the prompts and incorporate styles that will enable you to create unique creations, giving you a creative boost to ideate faster visually.

Bringing fictional concepts to life

I have also included some thumbnails and mentioned specific details of an art style/movement that you can discover. Enjoy!

Fictional concepts	

A futuristic transportation system – an extremely detailed, inked drawing in orange	A surreal landscape with floating islands – extremely detailed and in the Rococo style

A stream of consciousness in a vast landscape	A close-up of a dragon warrior, sliding down a mountainside packed with snow and steam coming off it

Table 8.2 – Sample Image thumbnails with prompts

Here are some that you can try on your own.

- A cybernetic cityscape where buildings are living organisms
- A steampunk-inspired underwater civilization with mechanical sea creatures
- A dystopian future where trees are made of circuit boards
- A post-apocalyptic landscape with skyscrapers overtaken by nature
- A whimsical forest inhabited by sentient mushrooms
- A surreal desert landscape with floating islands made of glass
- An alien marketplace bustling with creatures of all shapes and sizes
- A magical library where books fly off the shelves
- A retro-futuristic space station orbiting a neon-colored planet
- A fairy-tale castle guarded by mechanical dragons
- A time-traveling train speeding through different historical eras
- A cosmic carnival with planets as Ferris wheels and shooting stars as fireworks
- A parallel universe where gravity works in reverse
- A celestial garden with flowers that bloom into stars
- A haunted mansion with doors that lead to different dimensions
- A whimsical village floating on clouds
- A neon-lit cyberpunk alleyway filled with holographic advertisements
- A secret underwater laboratory researching mythical creatures
- A labyrinthine city with streets that rearrange themselves
- A robotic jungle where vines are made of cables
- A cityscape where skyscrapers are shaped like giant animals
- A hidden cave adorned with glowing crystals
- A magical circus with performers who can manipulate elements
- A post-apocalyptic wasteland inhabited by mutated creatures
- A futuristic metropolis where hover cars zip through glass tunnels
- A fantastical underwater kingdom ruled by mermaids and mermen
- A space colony built within the rings of a gas giant

- A cybernetic forest with trees that emit holographic leaves
- A mystical mountain peak shrouded in mist and guarded by mythical beasts
- A cityscape where buildings are powered by giant gears
- An enchanted waterfall cascading into a pool of liquid light
- A hidden city within a massive tree trunk
- A parallel dimension where reality is composed of geometric shapes
- A carnival on the moon with gravity-defying attractions
- A gothic castle atop a floating island in the sky
- A post-apocalyptic city reclaimed by nature, with vines overtaking skyscrapers
- A cosmic library containing books that hold the secrets of the universe
- A futuristic cityscape where people traverse on flying carpets
- A dystopian society where emotions are traded as currency
- A mystical forest with trees that whisper secrets to travelers
- An intergalactic marketplace bustling with alien traders
- A celestial observatory where stars are studied like artwork
- A cyberpunk metropolis where androids and humans coexist
- A hidden underwater civilization illuminated by bioluminescent creatures
- A fantastical garden with flowers that bloom into creatures of myth

Each of these prompts can inspire a wide range of artistic interpretations, showcasing different styles and techniques across various mediums.

Incorporating photographic techniques

One area that I am greatly interested in is generating photorealistic images using photographic techniques, which can help you ideate composing better-looking camera shots and is a great learning tool.

Let's check out the following examples.

Photo realistic images	

A macro shot of a flower petal with dew drops	A double exposure of a person and a natural landscape

A dramatic black and white landscape with strong contrast	Food photography – filet mignon medium rare with melting butter on top, Depth of Field, Bokeh, DOF, Tilt Blur, Shutter Speed 1/1000, F/22

Table 8.3 – Sample images generated using photographic techniques in the prompt

Here is a list of visual prompts you can try out in Adobe Firefly's Text to image:

- A high-key portrait against a white background
- A long exposure of city lights at night
- A silhouette of a person against a sunset

- A still life of vintage objects arranged on a wooden table
- Infrared landscape photography of a forest
- High-speed capture of a water droplet splashing
- A tilt-shift photograph of a bustling city street
- A light painting with colorful streaks against a dark background
- Minimalist architecture with strong geometric shapes
- Candid street photography capturing everyday life
- A **high dynamic range** (**HDR**) landscape of a mountain range
- A portrait using the Rembrandt lighting technique
- An abstract close-up of textured surfaces
- Motion blur of a moving vehicle on a highway
- A double exposure of a person and urban graffiti
- A vintage-style portrait with sepia tones
- Time-lapse photography of clouds moving across the sky
- Conceptual photography illustrating loneliness
- A high-speed photograph of a bursting balloon filled with paint
- Surreal levitation photography with a floating subject
- Reflections of city lights on water at night
- A composite image combining elements from different locations
- A portrait using the split lighting technique
- Abstract architecture with creative angles
- Multiple exposures of a dancer in motion
- A macro shot of a butterfly resting on a flower
- A minimalist composition with negative space
- Frozen motion of a water splash
- Urban decay photography in an abandoned building
- Bokeh photography of city lights at twilight
- A still life of colorful fruits arranged in a pattern
- A double exposure of a person and a forest landscape

- A high-key portrait with a bright, airy feel

- A low-angle shot of a skyscraper against a clear blue sky

- Vintage-style street photography in black and white

- A slow shutter speed capture of car lights on a winding road

- Abstract macro photography of water droplets on glass

- A portrait using the loop lighting technique

- A symmetrical composition of architectural details

- An intentional camera movement creating an abstract image

- A high-speed capture of a popping balloon filled with glitter

- A still life of old books and a quill pen on a desk

- A double exposure of a person and a starry night sky

- High-contrast black and white street photography

- A reflection of a person in a puddle after rain

I hope you enjoy checking out this list of prompts that you can use as you generate images with photorealistic techniques. In the next section, we'll check out how we can add more layers of complexity to our prompt by adding some recommended art styles.

Using art styles to create unique images

Art styles have been a cornerstone of expression for humans for centuries, and with the power of generative AI in Adobe Firefly, you can learn art styles faster by using Text to image to devise ideas quickly.

The following list is a series of suggestions to which you can add styles as you craft your prompts. It also inspired me to experiment by combining styles and, ultimately, made me appreciate art as it evolves over time.

You can mix and match in your next prompt using the art styles included in this list:

- Ancient art

- Folk art (all times)

- Renaissance (1350–1620)

- Baroque (1600–1750)

- Rococo (1740–1770)

- Line art (1940–1950)

- Geometric (900–700 BC)
- Pointillism (1880)
- Cubism (1880–1970)
- Minimalism (1950–1960)
- Romanticism (1800–1900)
- Fantasy (all periods)
- Art nouveau (1890–1914)
- Ukiyo-e (1615–1868)
- Impressionist art (1870s–1880s)
- Post-impressionism (1886–1905)
- Realism (1840–1848)
- Surrealism (1930–1940)
- Expressionism (1905–1920)
- Pop art (late 1950s)
- Modern art (1860–1970)
- Figurative (all periods)
- Abstract (1900–present)
- Pop art (mid–20th century)
- Graffiti and contemporary art (late 20th century–present)

I hope you will have a great time scrolling down memory lane with this list. My favorites are the art styles Rococo, art nouveau, and Ukiyo-e.

In the next section, let's discover where we can get assistance and ask for Adobe-related help.

Engage and get help with the Adobe Community

Need additional help? You can get in touch with the official Adobe Discord channels using this link: `https://adobediscord.com/`. It offers lots of opportunities to get assistance and guidance from industry practitioners by asking questions directly of the Adobe product managers and product community experts, who will surely offer you a helping hand.

You can join the Adobe Community Forum website using this link: `https://community.adobe.com/`. You can easily search for me by my name and I'm happy to connect. I always welcome a hard question that makes me learn more. Get yourself involved – it's as easy to sign up, using your Adobe ID, and then you can interact with people to discuss your concerns and feedback for Adobe apps. Lastly, you can check out a Facebook local community that we created, focusing on Adobe Firefly, at this link: `https://www.facebook.com/groups/521829953442730/`.

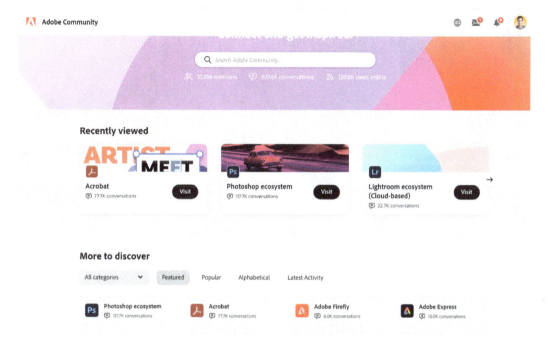

Figure 8.10 – The Adobe Support Community website

Final words

Congratulations on finishing this last chapter of the book. We covered a lot, from Firefly-related partnerships to the upcoming features for it, such as Text to Texture (Sampler) and Generative Background (Stager (Beta)) in Adobe Substance 3D.

We provided you with techniques to caption images that will help you do reverse prompt engineering, as well as a lot of resources that you can use and experiment with via the cheat sheet.

Thank you for completing this journey through the book with me; I appreciate your time and look forward to seeing you create images that will merge creativity and technology for the better.

I offer my final words in my native language (Filipino):

Maraming salamat at ipagpatuloy natin ang pagiging malikhain!

(Thank you, and let's keep on being creative!)

Index

T

packtpub.com

Subscribe to our online digital library for full access to over 7,000 books and videos, as well as industry leading tools to help you plan your personal development and advance your career. For more information, please visit our website.

Why subscribe?

- Spend less time learning and more time coding with practical eBooks and Videos from over 4,000 industry professionals
- Improve your learning with Skill Plans built especially for you
- Get a free eBook or video every month
- Fully searchable for easy access to vital information
- Copy and paste, print, and bookmark content

Did you know that Packt offers eBook versions of every book published, with PDF and ePub files available? You can upgrade to the eBook version at packtpub.com and as a print book customer, you are entitled to a discount on the eBook copy. Get in touch with us at customercare@packtpub.com for more details.

At www.packtpub.com, you can also read a collection of free technical articles, sign up for a range of free newsletters, and receive exclusive discounts and offers on Packt books and eBooks.

Other Books You May Enjoy

If you enjoyed this book, you may be interested in these other books by Packt:

Unlocking the Secrets of Prompt Engineering

Gilbert Mizrahi

ISBN: 978-1-83508-383-3

- Explore the different types of prompts, their strengths, and weaknesses
- Understand the AI agent's knowledge and mental model
- Enhance your creative writing with AI insights for fiction and poetry
- Develop advanced skills in AI chatbot creation and deployment
- Discover how AI will transform industries such as education, legal, and others
- Integrate LLMs with various tools to boost productivity
- Understand AI ethics and best practices, and navigate limitations effectively
- Experiment and optimize AI techniques for best results

Generating Creative Images With DALL-E 3

Holly Picano

ISBN: 978-1-83508-771-8

- Master DALL-E 3's architecture and training methods
- Create fine prints and other AI-generated art with precision
- Seamlessly blend AI with traditional artistry
- Address ethical dilemmas in AI art
- Explore the future of digital creativity
- Implement practical optimization techniques for your artistic endeavors

Packt is searching for authors like you

If you're interested in becoming an author for Packt, please visit authors.packtpub.com and apply today. We have worked with thousands of developers and tech professionals, just like you, to help them share their insight with the global tech community. You can make a general application, apply for a specific hot topic that we are recruiting an author for, or submit your own idea.

Share Your Thoughts

Now you've finished *Extending Creativity with Adobe Firefly*, we'd love to hear your thoughts! Scan the QR code below to go straight to the Amazon review page for this book and share your feedback or leave a review on the site that you purchased it from.

https://packt.link/r/1-835-08428-1

Your review is important to us and the tech community and will help us make sure we're delivering excellent quality content.

Download a free PDF copy of this book

Thanks for purchasing this book!

Do you like to read on the go but are unable to carry your print books everywhere?

Is your eBook purchase not compatible with the device of your choice?

Don't worry, now with every Packt book you get a DRM-free PDF version of that book at no cost.

Read anywhere, any place, on any device. Search, copy, and paste code from your favorite technical books directly into your application.

The perks don't stop there, you can get exclusive access to discounts, newsletters, and great free content in your inbox daily

Follow these simple steps to get the benefits:

1. Scan the QR code or visit the link below:

    https://packt.link/free-ebook/9781835084281

2. Submit your proof of purchase.

3. That's it! We'll send your free PDF and other benefits to your email directly!